The Difficult Road to Mars

   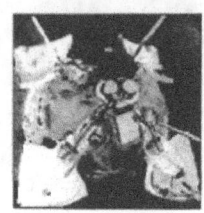  

A Brief History of Mars Exploration in the Soviet Union

By V.G. Perminov

A Joint Publication of the
NASA History Division
Office of Policy and Plans
and Office of Space Science

MONOGRAPHS IN AEROSPACE HISTORY
Number 15
July 1999

National Aeronautics and
Space Administration
Headquarters
Washington, DC 20546

NP-1999-06-251-HQ

V.G. Perminov was the leading designer for Mars and Venus spacecraft at the Lavochkin design bureau in the Soviet Union during the early days of Mars exploration. Here, he recounts the hectic days and urgent atmosphere in the Communist bureaucracy to design and successfully launch a Mars orbiter, a Mars lander, and a Mars rover. The goal was to beat the United States to Mars. The author's account gives, for the first time, the personal feelings of those managing the projects.

The first project was begun in 1959. During the next 15 years, the United States had put humans on the Moon, and the Soviet Union had put a cosmonaut in space and circled the Moon with a satellite. However, sending a spacecraft to a distant planet and having it enter an unknown atmosphere and land on a poorly known surface was an undertaking of a different magnitude. There were many lessons to be learned and many expensive failures. But with each new failure, new experience was gained, and with each successive attempt, the goal was closer.

In October 1960, with Project 1M, two spacecraft were launched, but the third stages of each rocket failed. In November 1962, the spacecraft Mars 1 was launched, but it fell silent at a distance of 106 million kilometers.

In March–April 1969, with Project M-69, there was an attempt to launch two spacecraft, but both failed on launch. In May 1971, with Project M-71, two spacecraft, Mars 2 and Mars 3, each with a lander, were launched. The lander for Mars 2 crashed on the surface of Mars. The lander of Mars 3 reached the surface, but its transmissions soon disappeared. However, the orbiters of Mars 2 and Mars 3 continued circling the planet for 8 months sending images to Earth.

In June 1973, Mars 4 and Mars 5 were launched. On Mars 4, the braking system failed, it therefore missed the planet. Mars 5 took images of Mars on a flyby. In August 1973, Mars 6 and Mars 7 were launched. Mars 6 was unable to receive commands after 2 months but, surprisingly, continued in an autonomous mode for another 5 months after landing on the Martian surface and sending back data. Mars 7 missed the planet.

During the mid-1970's, there were attempts to develop a program to return Martian soil to Earth. That program proved to be impractical.

In July 1988, the spacecraft Phobos 1 and Phobos 2 were launched to explore the Martian moon Phobos. Phobos 1 did not reach its destination. Phobos 2 successfully entered the Martian orbit, but at 150 kilometers from Phobos, it lost solar power and became silent. In November 1996, the spacecraft Mars 96, with an orbiter, four landers, and 22 scientific instruments, was launched. Because of onboard computer and upper-stage booster malfunctions, the Mars 96 spacecraft failed. This is the last spacecraft reported by the author.

In spite of numerous failures, the technical and scientific achievements during the Mars exploration effort were invaluable. The scientific results are broadly discussed in western literature, and technical knowledge has been advanced.

This translation was made by Dr. Katherine A. Nazarova for the East West Space Science Center of the University of Maryland.

Lev M. Muhkin
Deputy Director
East West Space Science Center
University of Maryland

CONTENTS

INTRODUCTION

Mars is the planet in our solar system thought to be most like Earth. The Martian period of rotation is 24 hours, 37 minutes, and its angle tilt with respect to its orbital plane is about 64.8 degrees, compared to 66.5 degrees for Earth. As a result, seasonal changes on Mars occur in the same manner as on Earth. Through a telescope, one can observe white polar caps on the Martian surface. As the summer approaches, the polar caps start to melt, and the Martian surface darkens with distance from the polar areas to the equator. Earth-based observations showed that near the Martian surface, the pressure was about 0.1–0.3 atmosphere, and at noon, the temperature near the equator was about 25 degrees Celsius. Because Mars has a very thin atmosphere, daily temperature variations on the Martian surface range up to 50 degrees Celsius. That is somewhat more than on the Earth's surface at the high elevations in the mountains, where the air is thin. Naturally, these similarities pose a question of life on Mars.

The idea of life on Mars appeared at the end of the 19th century after the Italian astronomer and director of the observatory in Milan, Giovanni Virginio Schiaparelli, discovered a network of fine lines, which he called "canals," on the planet's surface. Also on the Martian surface, Schiaparelli observed large dark areas, which he called "oceans." Dark areas of smaller size he named "lakes," and light yellow areas he named "continents."

The discovery of Schiaparelli attracted the attention of many astronomers. Using powerful telescopes, they managed to discover on the Martian surface many canals that always linked seas or lakes. In 1906, assuming that the canals on Mars did exist, American astronomer Percival Lowell put forward a theory that attempted to explain their origin. According to this theory, the canals were built by Martians to transport water from polar to arid areas. Schiaparelli and Lowell observed that the Martian surface changes with the seasons and suggested that this may be related to vegetation.

In the spring and summer, some areas of the Martian surface darken and acquire a greenish-blue hue. In autumn and winter, the same areas acquire a yellowish-brown hue. The best time to observe these changes is when the white polar caps start to melt. At this time, the Martian dark areas remind one of Earth's moist soil.

The change in the color of the surface was the greatest Martian mystery. In 1953, Soviet astronomer G.A. Tikhov tried to explain the color change. He pointed to the similarity between the reflection spectra of some areas on the Martian surface and moss on Earth's surface that grows in the dry and cold environment of the Pamir mountains at elevations of more than 6,000 meters. In 1953, American scientist H. Strughold speculated on the existence of primitive vegetation on Mars. The attitude of Soviet astrophysicist I. Schklovskiy was even more provocative. To explain the anomalous trajectory of the Martian moon Phobos, he suggested that Phobos was a hollow sphere. The proponents of intelligent life on Mars were delighted with this idea.

If primitive or intelligent life forms had been discovered on Mars, that would be of crucial importance for understanding the evolution of Earth and the universe. Certainly, the country that first detects life forms on Mars will be highly recognized and honored internationally. This ambitious goal was the main reason for the long competition between the Soviet Union and the United States. It is worth noting that the Soviet Union made a valuable contribution to this study. In spite of some setbacks, Soviet scientists and engineers made a large effort for the exploration of Mars.

The first stage of Mars exploration is finished. Surprisingly, vegetation, canals, and traces of intelligent life have not been found. However, dried-up courses of waterways have been observed. What happened? Why did water disappear? Did primitive or intelligent life exist in the past, or does it exist now on Mars?

Today, we cannot answer these questions. It now seems like Mars is a lifeless desert. On the other hand, we know that in Earth's deserts, archeologists dig up cities that flourished in the past but were neglected by ancient people, and they are now covered with sand.

In 1971, the largest dust storm ever registered by astronomers covered the whole Martian surface. For a few months, hundreds of millions of tons of dust were suspended in the Martian atmosphere. As a result, one could not observe the Martian surface. Nevertheless, one cannot rule out that cities covered by sand may exist in the Martian deserts. That would be evidence of an ancient Martian civilization that disappeared or moved to other planets.

Perhaps Martians once arrived on Earth and left some evidence of their visit. Perhaps they built huge runways in South America, maybe constructed a chemically pure iron column in India (now chemically pure iron can be produced only in the laboratory), and built the mysterious Egyptian pyramids.

It is also possible that the English fiction writer D. Swift managed to find and decipher records that Martians left on Earth. Based on these records, long before the Martian moons were discovered, Swift predicted that Mars had two satellites. One of them he named Phobos (fear), the other he named Deimos (horror), and rather precisely predicted the parameters of their orbits.

Possibly, the next generations of Mars explorers will clarify these questions. In particular, this book is written to preserve the record of the events of the first difficult road to Mars.

1.1 Project 1M

Ballistic rockets, which are able to carry heavy payloads, opened the way to interplanetary automatic spacecraft. For many centuries, investigations of other planets were limited to observations from Earth with telescopes at distances of tens of millions of kilometers from the planets. But generations of scientists dreamed of observations close to the planets. With the development of interplanetary automatic spacecraft, their dream was transformed into reality. To approach the planets, a spacecraft should fly in the vast regions of space for many hundreds of millions of kilometers. The conditions of interplanetary space were unknown and were described only as scientific hypotheses.

The first Martian spacecraft started to be developed in 1959 in the Experimental Design Bureau No. 1 (OKB-1) under the supervision of the Chief Designer and Academician S.P. Korolev. The preliminary design of the spacecraft included three major objectives:

1. To investigate interplanetary space between Earth and Mars

2. To study Mars from a flyby trajectory and to obtain images of its surface

3. To check the ability of onboard instruments to operate during the long flight in space and to provide radio communication from large distances

The scientific part of Project 1M was under the supervision of Academician M.V. Keldysh. At that time, he was Vice-President of the Academy of Sciences of the Soviet Union. Teams from different institutes of the Academy submitted scientific proposals. After close examination, a decision was made to put the following scientific instruments on the spacecraft 1M:

• Magnetometer

Top:
G.Yu Maksimov, the main designer

Bottom:
A.G. Trubnikov, responsible for the flight program and the spacecraft logic

Documentation for automatic Martian spacecraft was developed in OKB-1 and kindly given to us by G.Yu Maksimov and A.G. Trubnikov.

- Radiometer
- Charged particle detector
- Micrometeorite sensor
- Photo-television camera (FTU)
- Spectroreflectometer for determining the CH band, which may indicate the existence of organic life on the Martian surface

All scientific instruments, except the FTU, were attached to the outside of the spacecraft. The FTU was placed in a sealed module together with other onboard instruments and was designed to make pictures of the Martian surface through a viewport. It was designed so that as soon as the sensor indicated that the Martian surface was illuminated by the Sun, the televising would be initiated. The spacecraft was equipped with a permanent solar orientation sensor, which controlled the solar illumination for charging the batteries by the Sun during the whole flight. The attitude of the spacecraft in space was corrected in its trajectory by the Sun-star sensor. The correction was performed by the binary liquid-propellant engine, which runs on dimethylhydrazine and nitric acid.

It was proposed that an 8-centimeter wavelength transmitter and a high-gain antenna with a diameter of 2.33 meters transfer the Martian images to Earth. Its orientation with respect to Earth was supposed to be maintained with the help of radio bearing, which was obtained during the rotation of the spacecraft around the solar tube set to a predetermined angle.

To send commands to the outbound trajectory and to telemeter information, a decimeter wavelength radio transmitter was used. This radio system operated with a high-gain antenna. Two-square-meter solar panels and silver-zinc batteries were used for the power supply.

In October 1960, two spacecraft, with payloads of 650 kilograms each, were launched. Because of the failure of the third-stage rocket, neither spacecraft entered the proper flight trajectory to Mars. However, the effort of designers of the first Martian spacecraft was not a waste of time. Like small children, who fall and get bumps on their head while they learn to walk, the designers learned their own valuable lessons and accumulated the experience required for developing more sophisticated spacecraft.

1.2 Project 2MV

In the spring of 1961, Korolev directed the design of a new multipurpose spacecraft for the exploration of

Top:
G.S. Susser, responsible for the spacecraft configuration

Middle: Figure 2.
The Same Spacecraft With the Lander

Bottom: Figure 3.
The Launch Vehicle Transported to the Launch Pad

Mars and Venus. This was called Project 2MV and provided the opportunity for the exploration of Mars and Venus not only with a flyby trajectory but with a lander vehicle as well. The main design and the number of instruments on board did not change. To accomplish this task, the spacecraft was divided into two parts: the multipurpose orbital module, which delivered the spacecraft to the planet, and the module with the scientific instruments and equipment used to study the planet from a flyby trajectory (Figure 1). If the lander is to be used, it should be attached to the orbital module and replace the module with the scientific instruments (Figure 2).

Much effort was given to increase the reliability and improve the characteristics of the spacecraft. The radio system located in the orbital module was supplemented with a meter range wavelength radio transmitter, which broadcast from an omnidirectional antenna. The radio transmitter duplicated the main radio channel in the nearest part of trajectory. In the scientific module, in addition to the 8-centimeter wavelength transmitter used to transfer the planet's images to Earth, an impulse transmitter in the 5-centimeter wavelength range was installed. To point the high-gain antenna toward Earth, it was proposed to use a solar-Earth sensor with a mobile solar tube instead the radio bearing.

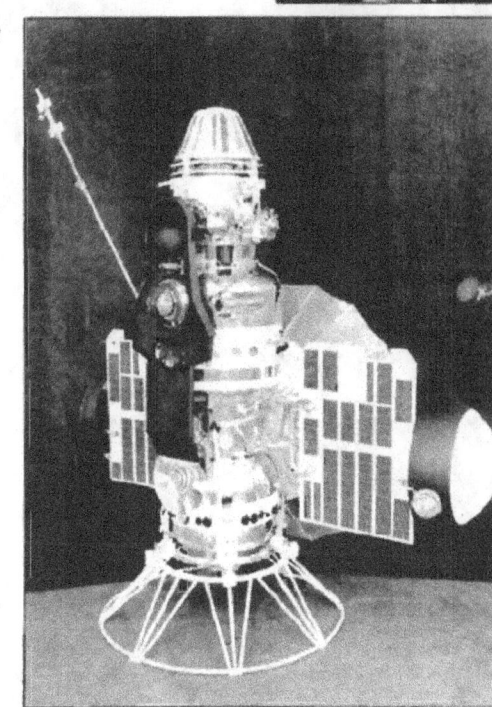

The solar tube was installed during the flight with a preset Sun-spacecraft-Earth angle. The area of solar panels was increased to 2.6 square meters. The silver-zinc battery was replaced by a cadmium-nickel battery with a capacity of 42 amp-hours. For temperature control, a binary gas-liquid system was used. For cooling and heating, liquid hemispherical coolers were utilized. In the upper part of the orbital module, a liquid propellant engine with a correction control system isolated by gimbal was installed.

On November 1, 1962, the four-stage rocket Soyuz (Figure 3) launched the spacecraft Mars 1 with a payload of 893.5 kilograms into a Martian trajectory. After the spacecraft and the fourth stage of the rocket were separated, the solar panels opened and a stable orientation of the spacecraft with respect to the Sun was maintained. However, the telemetered information was discouraging. It revealed that one

of the valves of the gas engines in the orientation system was leaking. Inevitably, this leakage would lead to the failure of the spacecraft. In this circumstance, the project managers decided to transfer the spacecraft to gyroscopic stabilization, which makes it possible to constantly illuminate the solar panels by the Sun.

Eventually, on March 21, 1963, the transmitters onboard Mars 1 fell silent. The last radio contact with the spacecraft was made when it was 106 millions of kilometers from Earth. Nevertheless, during the flight of Mars 1, the data on the characteristics of interplanetary space between Earth and Mars at the distance of 1.24 astronomical units were transmitted to Earth.

On November 30, 1964, the next spacecraft, Zond 2, designed for Mars exploration was launched in an interplanetary trajectory. In contrast to Mars 1, the radio system on Zond 2 did not include 8-centimeter and meter wavelength range transmitters. Six experimental plasma jet engines had been installed in the spacecraft. With a command from Earth, they could be used instead of the gas engines to control the motion of the spacecraft around its center of gravity. Unfortunately, Zond 2 did not fulfill its mission because the solar panels did not open entirely.

2.1   Korolev: Triumph and Tragedy

In the late 1950's and the early 1960's, OKB-1, headed by Korolev, made a significant contribution to the development of space technology. For a short period of time, the following outstanding projects were developed and utilized:

- 1956—A powerful intercontinental ballistic rocket was built.
- 1957—An Earth satellite was launched.
- 1959—Lunar spacecraft were launched.
- 1960—An unmanned spacecraft was launched into an Earth satellite orbit.
- 1961—The flight of Yuri Gagarin was made.
- 1962—The automatic spacecraft Mars 1 made the first steps into vast regions of the space.

In 1956, after the successful completion of the testing of the rocket R-7, designed by Korolev and known in mass media as Sputnik, the realization of these projects gave the Soviet Union the leadership in space exploration. The legendary R-7 rocket and its modifications were reliable, powerful, and able to deliver a payload of 6,000 kilograms in orbit.

In addition to the above-mentioned projects, the designers from OKB-1 developed military missiles, communications satellites, reconnaissance satellites, and remote-sensing satellites. Naturally, Korolev and his colleagues were too busy to concentrate on the main project—a manned flight to the Moon. The Lunar project required the design of the powerful rocket N-1, which would be able to launch a payload of 100 tons and a lunar spacecraft into lunar orbit.

At that time, the tremendous success of the Soviet Union in space exploration put the United States far behind. To reestablish the prestige of the United States, President Kennedy announced a lunar Apollo project to be a national priority. The unprecedented race between the two powerful countries began. To be able to concentrate on the lunar project, Korolev ordered a number of interesting projects to be transferred to other

Top:
S.P. Korolev, main designer of OKB-1

Bottom:
G.N. Babakin, main designer of the Lavochkin design bureau

organizations. At the same time, he was concerned that these projects would be developed without his supervision. The fact that these organizations were headed by the former main designers of OKB-1 made him feel better.

- D.I. Kozlov moved to Samara (Kuibyshev). He was directed to develop and modify the rocket R-7, and the reconnaissance satellites as well.
- In the town of Miass (in the Ural mountains), under the supervision of V.P. Makeev, military missiles were designed.
- In Krasnoyarsk, a new design bureau headed by M.V. Reshetnev continued to develop communications satellites.

Soon these organizations would become big institutions and make significant contributions in space technology. Korolev continued to design automatic spacecraft for the exploration of the Moon, Mars, and Venus. He did not want to give up his dream of flights to the other planets, especially when victory seemed to be so close. His faith was based on the fact that the new spacecraft design successfully corrected all previous problems and mistakes. Unfortunately, his faith was not justified. Because of the confusion in the technical documentation, the programs of the flights had not been performed completely. This negligence was the result of the fact that the team of designers was involved in too many activities and could not concentrate on the one project.

In April 1965, Korolev issued instructions to continue the design of the lunar and planetary spacecraft in the Scientific Production Association (NPO), named after S.A. Lavochkin (Lavochkin design bureau). At this time, the project was headed by G.N Babakin, the main designer of the Lavochkin design bureau.

The design team of the Lavochkin design bureau had extensive experience in the development of automatic aircraft, particularly intercontinental ballistic cruise missiles and unmanned airplanes. Few of these designers had any experience in the design of spacecraft.

By the end of 1959, Lavochkin had organized a team of 15 young designers who were already well known for using new techniques and instructed them to design a space aircraft. The idea was that the aircraft should take off, enter a lunar orbit, and return to land at an airport. Lavochkin told us, "This assignment is intricate and will take a lot of time. That is the reason why I decided to organize the team of young designers so you would have time to finish the project."

For the project to make progress, Lavochkin needed to hire more designers. Unfortunately, in June 1960, after Lavochkin passed away, the team was dismissed and the project for the design of a space aircraft was discontinued. To continue to be employed, the designers from the Lavochkin design bureau should recall how to design spacecraft and combine this knowledge with the skills acquired in the design of unmanned aircraft. Also, they wanted to use the valuable experience of the designers of OKB-1.

OKB-1 was considered a very good school for the specialists in systems analysis, and their experience was invaluable. On the other hand, the design team, recently combined with the engineers from other organizations, consisted of relatively young and inexperienced designers. Therefore, the qualification of the designers from the Lavochkin design bureau, who were considered experts in developing systems with strict weight limitations, was unquestionable. In a few months, our engineers designed the new orbital modules for the spacecraft named Mars and Venus. As a big surprise to the designers from OKB-1, the weight of the orbital module was decreased by a couple of tens of kilograms. The OKB-1 specialists did not believe that would be possible.

The first and last time Korolev visited the Lavochkin design bureau was in July 1965. At that time, with the help of designers from OKB-1, we modified the lunar and Martian spacecraft. The results of this revision were shown on posters that were hung on the walls in the office of G.N. Babakin. We discussed the problems of lunar spacecraft. The next lunar spacecraft was to be launched at the beginning of 1966.

I stood close to the poster in which the Martian spacecraft with the lander in the shape of a plate was shown. In the same poster, the characteristics of the parachute landing on the Martian surface were depicted. At that moment, Korolev and Babakin entered the office. One could see that Korolev was thinking about something quite important to himself. He came close to the poster of the Martian spacecraft and thought for a while. After that, with a gloomy look and resting his chin on his hand, he, quietly, not expecting to be overheard, said, "The landing should be performed by the engines, without parachutes." In response, I timidly reminded Korolev that Mars possesses an atmosphere. Glancing at me, he turned around and approached the table. Being the chairman of the meeting, he invited everybody to take a seat.

Various experts who represented many organizations participated in the meeting: Babakin—the main designer of OKB-1, I.N. Lukin—the head of the Lavochkin design bureau, V.E. Izhevskiy—deputy to the main designer, I.A. Skrobko—the leading designer of the lunar spacecraft, V.G. Perminov—the leading designer of the Mars and Venus spacecraft, M.I. Tatarintsev—the head of the design bureau, M.K. Rozhdestvenskiy—the deputy director of the design bureau in the fields of aerodynamics and thermal control, D.K. Brontman—the head of the design for systems analysis, R.S. Kremnyov—the secretary of the Communist Party Committee of the Lavochkin design bureau, and G.Yu. Maksimov—the head of the department of OKB-1 for the designing of lunar and planetary spacecraft.

Korolev spoke briefly. He emphasized how hard the team was working on the lunar project and the frustrating mistakes discovered during the flights of the spacecraft. In addition, he said that it is not feasible to continue to work in the same manner and suggested transferring the promising project to the Lavochkin design bureau. His final remarks are still in my memory. He said, "I hand over to you the most

Top:
I.N. Lukin, director of the Lavochkin design bureau

Middle:
V.E. Izhevskiy, deputy to the main designer

Bottom:
I.A. Skrobko, lead designer of the lunar spacecraft

Top:
V.G. Perminov, lead
designer of the Mars and
Venus spacecraft

Middle:
M.I. Tatarintsev, head of
the design bureau

M.K. Rozhdestvenskiy,
deputy director of the
design bureau

Bottom:
R.S. Kremnuev, secretary
of the Communist Party
Committee of the Lavochkin
design bureau

valuable possession—my dream. I expect you to work hard. But if my faith is not rewarded I'll do as Taras Bulba[2] once said: 'I gave you life and I'll take your life.'"

2.2 Challenges en Route to Mars

The decision was made that the engineers from OKB-1 should finalize the design of the spacecraft Venus 2 and Venus 3 for planetary exploration with a flyby trajectory and with a lander. Simultaneously, the engineers from the Lavochkin design bureau were directed to design spacecraft for further planetary exploration.

Only 1.5 years remained until the next astronomical window, when the launch of the spacecraft to Mars would be feasible. During this period of time, we were instructed to review the design of the previous spacecraft, develop a design for a new spacecraft, and fabricate and launch it to Mars. Naturally, we did not want to spend money on experiments that had already been done with the spacecraft Mariner 4. One of the main goals of our project was not to repeat the experiments of Mariner 4. It is worth noting that Mariner 4 transmitted only preliminary data to Earth from a flyby trajectory. Still a lot of efforts would be required to understand the origin of Mars.

At first, we expected that designers from OKB-1 would share their expertise with us. Babakin identified a number of designers from each design team and the key personnel according to their qualifications. Specifically, our team was responsible for the development of the spacecraft design. Also, we were expected to be familiar with the technological achievements of the main designers in other organizations.

Soon we concluded that the project and system management structures in OKB-1 and in the Lavochkin design bureau were fundamentally different. Project management responsibility in OKB-1 assumed:

1. Preliminary design and configuration of the spacecraft

2. Definition of the main systems and their parameters

3. Development of the flight program and its rationale

4. Supervision and management during the whole flight

The technical part of the project was supervised by the head of the design division. The main designer did not participate in the project development and took responsibility only after the divisions of OKB developed and issued the technical documentation. It was certainly a hierarchic management structure.

---

[2] Taras Bulba is a fictitious character in the novel by Gogol, *Taras Bulba*. In this novel, Taras Bulba kills his son for betraying the country.

Traditionally, in the Lavochkin design bureau, the so-called matrix management structure was used. During the development of the preliminary design of the spacecraft, the main managers of the project design bureau were temporarily in charge of qualified personnel from other design bureaus. Each professional was in charge of his own system and, as a member of the temporary team, participated in the development of the preliminary design. After the preliminary design was completed, the qualified personnel returned to their design bureaus and continued to develop and issue technical documentation for other systems. From the beginning of the project, the main designer was in charge of technical management.

The previous projects developed in OKB-1 and the Lavochkin design bureau demonstrated the efficiency of both management structures. We believed our system to be more effective because, for a long time, our team successfully utilized it, and we did not see any reason to change.

Our specialists were successfully completing a training program because in most cases the technical approaches were the same as before. However, some problems were discovered in the operation and attitude control systems of the spacecraft. In our previous projects, the operation of the craft during the whole flight was accomplished with gyroscopes. Gyroscopes were corrected with the radio system.

The operation and attitude control of the spacecraft during the whole flight was maintained with the Sun, star, and Earth sensors. Gyroscopes were turned on for a short period of time when the correction of the trajectory or the damping of the angular velocity of the spacecraft was performed.

The project managers did not believe that our personnel in the short time span would be able to comprehend all the nuances of this new design. Therefore, they decided that OKB-1 would continue to develop the operational and attitude control systems. The results would be transferred to the Lavochkin design bureau, where they would be used for designing spacecraft based on technology developed in the Venus 2 and Venus 3 projects. With small modifications, these spacecraft could be used to measure the parameters of the Martian atmosphere with a flyby trajectory and with a lander if it would replace the scientific module.

The flyby option did not give any possibility of obtaining new data on the Martian atmosphere and was excluded. To study the Martian atmosphere, it was necessary to increase the weight of the lander midsection because the data from Mariner 4 showed that near the Martian surface, the pressure was less than 0.09 atmosphere. However, the lander developed in OKB-1 was designed for a pressure of 0.1–0.3 atmosphere. We tried to determine whether it would be feasible to simultaneously increase the weight of the lander midsection and provide a soft landing to the Martian surface. Eventually, we decided that the change in the configuration and the number of the onboard instruments would not give us the opportunity to increase the weight of the lander midsection.

The specialists from OKB-1 based this decision on a careful consideration of this problem. The experts on the temperature control system suggested that the increase in the lander midsection weight could be achieved if the gas-liquid temperature control system would be replaced by a gas-gas system.

However, at that time in the Soviet Union, the thermal vacuum chambers with a Sun imitator were not available, and the replacement of a reliable system, already checked in real flight, with a new one was very risky. Eventually, the reconfiguration of the design of the orbital module was completed. As a result, a decrease in the thickness of module walls and in the weight of instrument frames was made. This was utilized to increase the weight of the lander.

Nevertheless, even after the weight of the lander was increased, its descent with the parachute in the Martian atmosphere took about 25 seconds. Taking into account that the rate of transmitting the information from the lander to Earth was 1 bits/sec, this time was definitely insufficient to obtain the reliable data on the Martian atmosphere. At the same time, a soft landing on the Martian surface would make it possible to significantly increase the time of transmitting information. Unfortunately, because of weight limitations, a soft landing was not feasible.

Thus the conclusion was reached that Mars exploration with a first-generation spacecraft had no future. In October 1965, the development of the first-generation spacecraft was discontinued, and the development of the second-generation spacecraft began.

2.3 The Second Generation of the Spacecraft

In 1965, the new powerful two-stage Proton rocket was launched for the first time. This rocket delivered in the predetermined orbit a payload of more than 12 tons. This was almost twice as much as Molniya, the recently modified three-stage R-7 rocket delivered.

Simultaneously, the development of a three-stage Proton rocket, which would be able to deliver in orbit a payload of 17–20 tons, was planned. The development of the second-generation spacecraft could be based on this launch vehicle. It was planned that the new multipurpose spacecraft would perform major experiments during the reconnaissance flights. The data obtained would be used to develop more advanced spacecraft, which would allow more detailed exploration of interplanetary space and the planets in our solar system.

Assuming that in 5–7 years the program of reconnaissance flights would be completed, I felt it was necessary to build a multipurpose versatile spacecraft that would be able to simultaneously explore Mars and Venus and solve the scientific and technical problems that appear during the flight. To reduce production costs, decrease the time required for developing and building the spacecraft, and enhance the probability of a successful flight, it was planned not to make significant changes in the design of the spacecraft and its onboard systems.

This approach was approved by Babakin and required the careful selection of experiments to be carried out by the future spacecraft. The ability to transmit a large quantity of information was analyzed as well. Soon, because of effective cooperation between scientists and designers, the definition of spacecraft parameters and the major objectives of the flight were successfully completed. This cooperative effort reminded one of the ascents of two tightly connected hikers to the top of an unknown steep hill.

On March 22, 1966, Babakin made the handwritten comments in the proposed document. Simultaneously, he justified the major issues proposed as the cornerstones for the successful development of the second-generation spacecraft for the exploration of Mars and Venus in the period from 1969 to 1973. These issues included:

1. The use of a three-stage Proton rocket for launching the spacecraft and booster block in a predetermined geocentric orbit.

2. The use of a descent-flyby and descent-orbital designs in the flight profile and increase of the weight of lander vehicles to provide a reliable landing and to place scientific instruments on the Martian surface.

3. The use of a universal propulsion system for the trajectory correction and for the launch of the spacecraft in the planet's satellite orbit with a pericenter of about 2,000 kilometers and an apocenter no more than 40,000 kilometers. The propulsion system was to be designed as a multipurpose module.

4. The use of the flyby orbiter or the planet's satellite to retransmit the information from the lander to Earth at a rate about 100 bits/sec.

5. The transmittal of scientific information from the spacecraft to Earth at a rate of about 4,000 bits/sec.

The following scientific problems were to be solved during the reconnaissance flights of the second-generation spacecraft.

A. The proposed missions for Mars exploration:
1. To measure the temperature, pressure, wind's speed, and direction on the Martian surface. To measure the chemical composition of the Martian atmosphere with position reference. It was proposed that in 1969 a lander would be used to acquire a pressure and temperature atmospheric profile. At that time, a soft landing was not planned.
2. To perform a soft landing at a chosen site and use the lander to obtain images of the Martian surface to study the relief and vegetation.
3. To measure the parameters of the Martian soil (composition, rigidity, and temperature).
4. To measure the radiation and the intensity of the magnetic field at the Martian surface.
5. To detect traces of microorganisms in the Martian soil.

6. To study the Martian upper atmosphere.
7. To compile a detailed radiothermal map of Mars.
8. To obtain from the flyby orbiter the Martian moons' images to define their shape, size, and albedo.
9. To get the images of the Martian surface from the orbiter to understand the nature of "seas" and "canals" and to acquire information on the seasonal changes on the Martian surface.

B. The proposed missions for Venus exploration:
1. To acquire data on Venus' atmospheric profile with altitude reference (temperature, pressure, composition, and illumination).
2. To study the chemical composition of the atmosphere near the surface and to detect microorganisms.
3. To make images of the surface of Venus using the camera installed in the lander.
4. To investigate the aggregation states and the mechanical properties of the soil.
5. To study the upper atmosphere.
6. To compile a detailed radiothermal map.
7. To get images of Venus using the cameras installed in the orbiter.
Besides the above-mentioned problems, the spacecraft would study the following characteristics of the interplanetary and near planetary space:
   (a) magnetic, electrical, and gravitational fields and the radiation environment
   (b) solar and space radiation
   (c) the meteorite environment

The launch of the first two spacecraft of the second generation, which were to explore Mars from their satellite orbit and with the probe device, was planned for the nearest launch window in March 1969. Only 3 years remained until this date. However, we believed that under the proper management and supervision, the problems could be solved.

Taking into account the proposed missions, our specialists started the spacecraft design. Unfortunately, because of the malfunctioning of the temperature control system, the spacecraft Venus 2 and Venus 3, as they approached the planet in February–March 1966, were not able to conduct the investigations planned.

In April 1966, the government decided that the next expedition to Venus would be in the following launch window of June 1967. This task seemed to be too ambitious. In 13 months, we were required to develop the new design, fabricate the spacecraft, perform all testing operations, and launch the spacecraft to Venus.

The Soviet Union was also competing with the United States, and leaders of our country were unwilling to concede defeat. Now all efforts were focused on building the spacecraft for Venus exploration.

3.1 The First Version of the Preliminary Design

The exhausting work of building the Venusian spacecraft was completed. On June 12, 1967, the spacecraft Venus 4 was launched into an interplanetary trajectory. After a short break for vacations, our specialists started to work on a new project for Mars exploration, which was scheduled to start in 1969.

The project was named M-69. Now the launch window to Mars was only 20 months away. The possibility of postponing the launch of the spacecraft until the next stage was not even considered. The designers, encouraged by the successful work on the Venus Project, were drawn into competition and were consumed with the desire to win. We tried to find a way to restrict the number of proposed missions; however, the main missions planned for 1969 remained unchanged. The main missions were:

1. Exploring Mars with an orbiter
2. Acquiring data of the Martian atmospheric profile with the entry probe

At that time, the designers from the Lavochkin design bureau were developing a third-generation spacecraft for lunar exploration. The primary component of this spacecraft was the propulsion system with four spherical fuel tanks connected by modules of cylindrical shape. Onboard instruments were located in the cylindrical modules. The technical documentation for the fuel tank was already developed, and we were looking into industrial opportunities for its fabrication.

For the Martian spacecraft, the engineers proposed the use of the design of the propulsion system, which was already used for the lunar spacecraft. Only a few modifications, related to the installation of the modules between the tanks, the replacement of some instruments, and a change of the sequence in which fuel was burned, were suggested. The Martian lander was attached to the spacecraft on the side (Figure 4). Exactly at this place, the lunar rover was attached to the lunar spacecraft. This configuration restricted the use of the spacecraft for further flights and contradicted major instructions. However, it allowed the spacecraft to be launched in 1969.

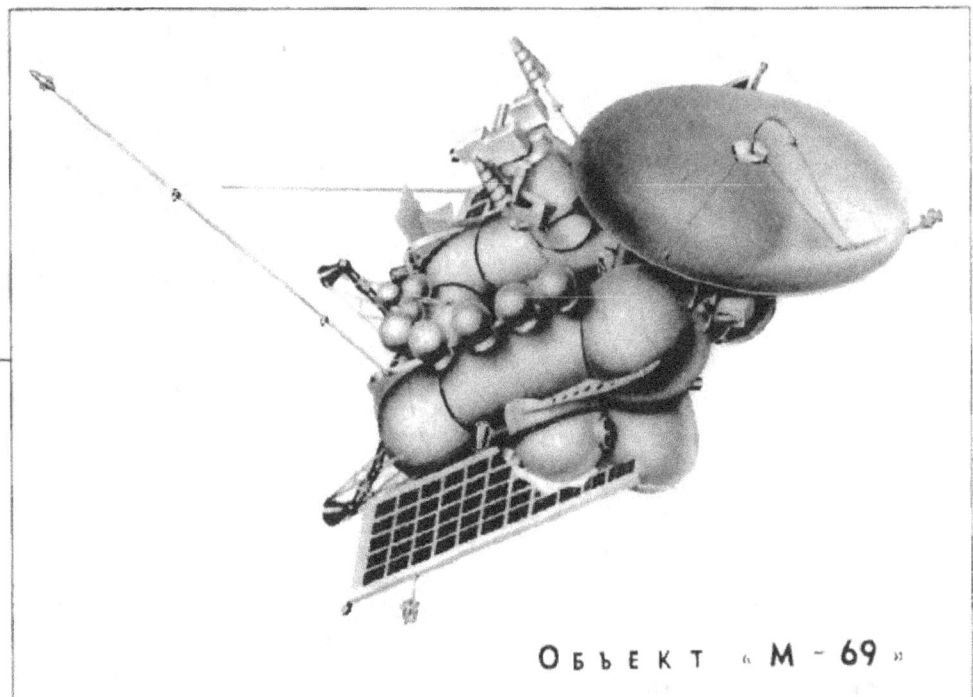

Обьект « М - 69 »

Figure 4.
The M-69 Spacecraft
(First Version)

In November 1967, Babakin approved the preliminary design of Project M-69. Two spacecraft were to be launched with the following purposes:

1. To obtain information on interplanetary space
2. To study the Martian environment
3. To check the performance of the new onboard systems and the reliability of new materials used for building the spacecraft
4. To obtain information that can be used during a soft landing on the Martian surface in 1971

It was proposed that a three-stage "Proton" rocket be used to launch the spacecraft and the booster block D in a circular orbit with an altitude of 200 kilometers. Utilizing the block D upper stage and the correction braking unit of the propulsion system, the spacecraft would be transferred to an interplanetary trajectory in two stages.

Descent orbital flight profiles were chosen. As the Mars encounter is approached, the lander would be separated from the spacecraft and the solid fuel engine would be used to transfer the spacecraft to a trajectory for a planetary encounter. The calculated angle at which the lander should enter the Martian atmosphere was in the range of 10–20 degrees.

During deorbiting, when the speed of the lander in the Martian atmosphere would be decreased to Mach 3.5, the parachute deployment would be initiated and data on the Martian atmosphere would be transmitted to Earth. The calculated altitude of the parachute system deployment would depend on the angle at which the lander entered the Martian atmosphere and would vary from 2.2 to 31.7 kilometers. In accord with that, the time of information transmission would change and vary from 30 to 900 seconds.

After the separation from the lander, the spacecraft would keep moving in its trajectory. After approaching the target point, the correction braking propulsion system would transfer a braking impulse equal to 1,750 m/sec. Depending on the real height of pericenter of the flyby hyperbolic trajectory, the spacecraft would enter an orbit of the Martian satellite with the following parameters: the height of the pericenter 2,000 ±1.5 kilometers, the height of apocenter would vary from 13,000 to 120,000 kilometers, period of rotation would vary from 8.5 to 12 hours, and the inclination of the orbit would vary from 35 to 55 degrees.

To prove the reliability of the spacecraft and its systems, we planned to focus our efforts on the following problems:

1. The study of the spacecraft characteristics during flight
2. The study of the aerodynamic characteristics of the lander and the reliability of the parachute system during lander deorbiting
3. To develop techniques for lander sterilization
4. To check the reliability of the elastic nonmetallic membranes designed for the displacement of the fuel from the tank

The feasibility of long-term storage of the elastic membranes in the fuel tanks had already been tested during the lunar project. Naturally, the flight time of the Martian spacecraft would be much longer. The test experiments should answer to the question on the reliability of membranes after the long-term storage in the fuel tank. The results of these experiments would determine the fate of the project.

3.2   The Second Version of the Preliminary Design

As the proposed study made progress, new disadvantages of the chosen spacecraft configuration were revealed. During flight, the moment of inertia of the spacecraft would have changed significantly because the consumed fuel would change the center of gravity. This problem prevented us from making a unified adjustment of the control system. In addition, because of irregular fuel consumption, the eccentricity of the engine thrust would be increased. As a result, there was a decrease in the predicted accuracy of the trajectory. At the same time, the range of angles at which the lander vehicle might enter the Martian atmosphere and limitations on the time required for the spacecraft to pass the target point would increase. Also, the temperature control system, which was designed to provide the appropriate temperature environment for instruments located in the three isolated modules, was becoming quite complicated.

However, a major problem was discovered during testing of the elastic membranes, which were designed for the displacement of the fuel from the tank. Testing of the membranes after a few months of storage showed that sometimes in fracture areas they were not hermetically sealed. Because of the shortage of time, we were unable to find the proper solution that would guarantee reliable sealed membranes.

We started to analyze the results of a study that was performed by the team of designers (V.I. Smirnov, A.Y. Fisher, and others). To start the engines in the weightless state, they attempted to develop the new system, which was composed of the main and supply tanks. They suggested that the lenticular-shaped supply tank with the metallic membranes and a valve that regulated the consumption of fuel should be placed in the main tank. To exclude the bubbles in the engine, the vacuum processed fuel from the supply tanks was used in starting the engine. Under acceleration, 6–8 seconds after the engine was started, the fuel was forced against the bottom of the main tanks, the bubbles floated up to the surface, and the fuel for the engine started flowing from the main tanks. Unfortunately, this bright idea could not be used in the design of an engine consisting of four tanks, because it led to an unacceptable eccentricity of the engine thrust.

After considering all the pros and cons and taking into account the opinion of the main designers, Babakin made an unexpected and risky decision. He decided to discontinue work on the existing design of the spacecraft and to start developing a new spacecraft design using an engine with supply tanks. At that point, only 13 months remained until the planned launch date of the spacecraft.

Surprisingly, the decision of Babakin did not depress the team. On the contrary, it seemed that the whole team got new inspiration. The pace of work, although quite impressive before, was increased further. In a relatively short time, the new preliminary and configuration designs of the spacecraft were developed, and the design documents were published.

Compared with the first version, the changes were crucial. Now the spherical fuel tank was located in the center of the spacecraft (Figure 5). An inner baffle divided the fuel tank into two tanks consisting of fuel and oxidizer. The lenticular-shaped supply tanks with the metallic membrane and the valve-switch were installed in each tank (Figure 5). Two hermetically sealed cylindrical-shaped modules with the instruments

Figure 5.
The M-69 Spacecraft
(Second Version)—
(1) parabolic high-gain
antenna, (2) lander,
(3) fuel tank, (4) solar
panels, (5) engine,
(6) nozzles of the
attitude system,
(7) radiator-cooler,
(8) viewports of the
photo-television camera
(FTU), (9) module with
the instruments,
(10) radiator heater,
(11) omnidirectional
antenna, (12) system of
astro navigation, and
(13) automatic inter-
planetary station M-69

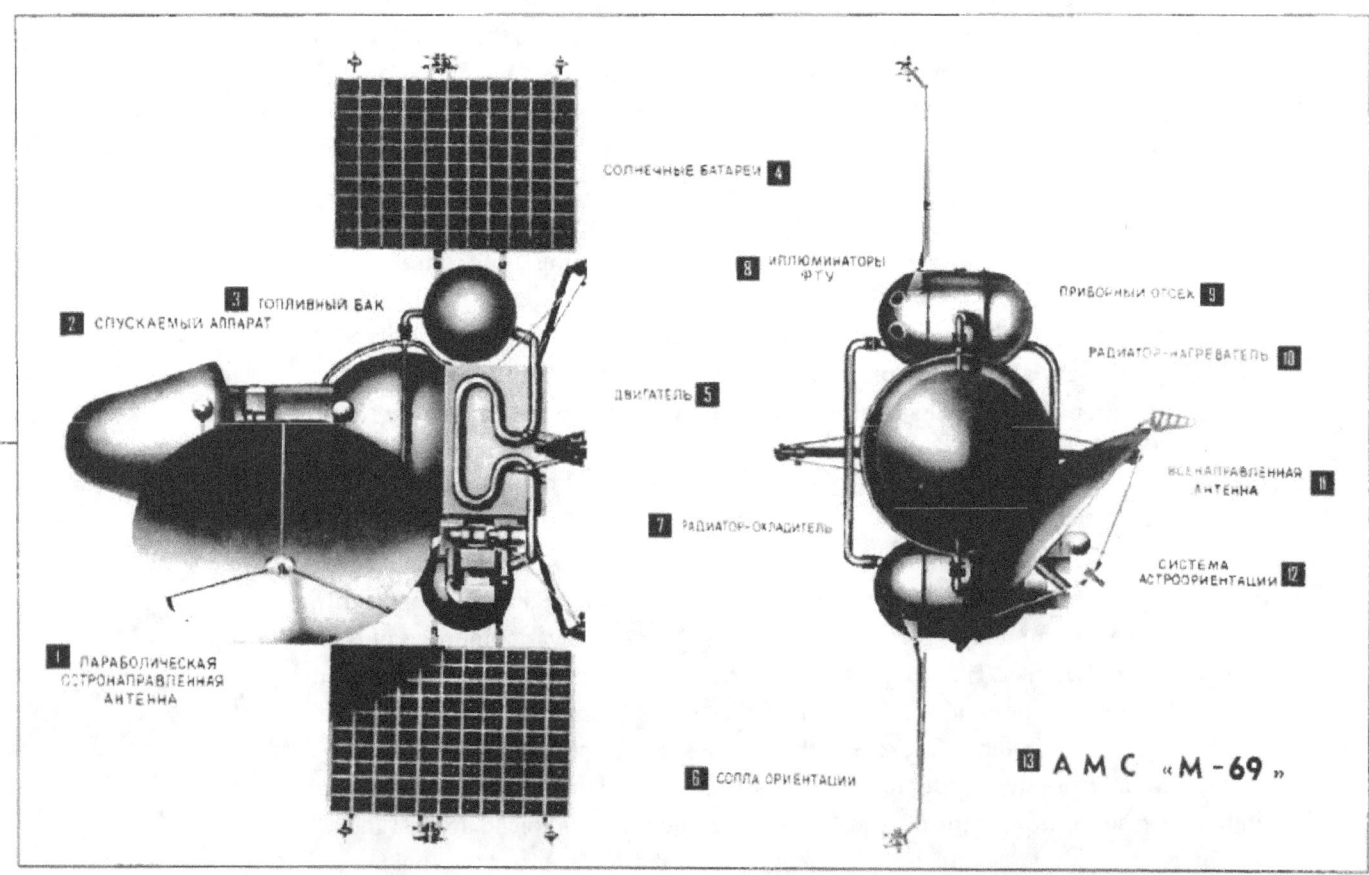

were attached to the spherical tank. The lander vehicle, in the shape of a headlight, was attached to the upper surface of the tank. The number of instruments and their parameters as well as the flight trajectory, excluding the final parts, did not change.

To increase the time of deorbiting of the lander vehicle on the parachute, it was decided not to separate the lander from the orbiter at the time when the spacecraft was approaching the planet, but to do that when the spacecraft entered the corrected orbit of an artificial Mars satellite. Because of the increased accuracy of the outbound trajectory, the error of pericenter definition for the initial orbit of this artificial Mars satellite was decreased from 2,000 to 1,000 kilometers. Simultaneously, the error of the time of flight definition was decreased from ±10 minutes to ±5 minutes. Then the maximum height of the apocenter of the initial orbit decreased to 70,000 kilometers, and the maximum period of rotation decreased to 65 hours. The proposed weight of the spacecraft was now 3,834 kilograms. This number included the weight of the lander vehicle (260 kilograms).

The following scientific instruments were to be installed on the spacecraft:
(a) Magnetometer
(b) Meteorite detector
(c) Low-frequency radiation detector
(d) Charged particles detector
(e) Cosmic ray and radiation belts detector
(f) Spectrometer of low-energy ions
(g) Radiometer
(h) Multichannel gamma spectrometer
(i) Mass-spectrometer H and He
(j) X-ray photometer
(k) Ultraviolet photometer
(l) Infrared Fourier spectrometer
(m) Three telephotometers with the focal distances of 35, 50, and 250 millimeters

To study the Martian atmosphere, it was planned to install the gas analyzer and the detectors of pressure, density, and temperature in the lander. The total weight of the scientific instruments installed on the spacecraft was 99.5 kilograms. This number included the weight of scientific instruments installed in the lander (15 kilograms).

The preliminary project required that the first spacecraft should be launched on March 24, 1969, and the second spacecraft on April 2, 1969.

3.3   The Government Decision Should Be Accomplished

A joint decree of the Communist Party Central Committee of the Soviet Union and the Council of Ministers of the Soviet Union assigned that the spacecraft for Mars exploration should be launched in 1969. To accomplish this task, the Commission of the Presidium of the Council of Ministers issued a special Military Industrial Resolution

Minister S.A. Afanasiev, head of the Ministry of the General Machine Building

(called a VPK Resolution). This resolution defined the program and the schedule of work of each organization that was involved in the building of the spacecraft and that performed the testing operations. The fulfillment of the VPK Resolution was strictly controlled. If it would not be accomplished, we could be in big trouble. On the other hand, its successful completion would be highly honored and awarded.

Babakin clearly understood how delicate the situation was. However, not wanting to be in trouble, he did not consider to propose that the government postpone the launch of the spacecraft to the next launch window. Recently, while the spacecraft Venus 4 was being fabricated, he become convinced that his teammates could work very efficiently round the clock and were able to solve extremely intricate scientific and technological problems in a very short period of time.

The ability of the team to work efficiently was developed by Lavochkin, who in the 1950's hired many graduates from the Moscow State Technical University (also known as Bauman Institute) and from the Moscow and Kazansky Aviation Institutes. Since that time, the young specialists added practical experience to their excellent knowledge of theory and became highly qualified professionals. They set the pace of work and felt like pioneers who discover new worlds.

However, the older generation also worked hard. Routinely, very late at night in the office of Babakin, the project managers tried to find the ways to solve the next problem. Many divisions worked round the clock. The schedule was tough, and it was common for the specialists and main managers to work at night. Because of the shortage of time, some jobs were scheduled to be completed at night, and if a problem appeared, it should be solved immediately. The chauffeur of a special car was ordered to quickly deliver the specialists to work at nighttime.

Quite often, I was awakened during the night and was taken to work from a warm bed. There were wild nights, when problems appeared in several divisions simultaneously. Then, my wife worked like the secretary and by telephone told where one can could find me.

Our Minister S.A. Afanasiev, one of the most talented leaders of the Soviet Union industry, helped us a lot. He was the head of the Ministry of the General Machine Building, which was in charge of the organizations with more than 1 million employees. It was a huge staff. Afanasiev did not like to procrastinate and made decisions quickly. He realized that to meet the schedule, some problems should be solved immediately. Sometimes in the course of the spacecraft fabrication, it was necessary to develop a new system or an instrument in an organization that did not belong to our ministry. Apparently, in the VPK Resolution, the new devices were not mentioned because the problem appeared after the resolution was issued.

Naturally, the main managers who were not mentioned in the VPK Resolution did not want to be in charge of new devices. In this situation, Afanasiev carefully considered how to accomplish this task with the organizations of our ministry. Only after Afanasiev was aware that it was not feasible, he asked for help from the minister of another branch. After his telephone call and in spite of the fact that our specialists were not very warmly welcomed by the staff of the other ministries, the job was usually finished on schedule. Afanasiev was highly esteemed, and his request was never rejected.

Afanasiev had been a big help in building the different spacecraft systems and parts. Learning that the experimental plant could not fabricate the parts of spacecraft on schedule, he ordered the main managers of the Urals and Siberian plants to manufacture the necessary spacecraft parts according to our technical documentation and to deliver them to the experimental plant. Usually, the Orenburgskiy plant fabricated the tank units, the Ust-Katavskiy plant fabricated the units for the automatic propulsion system, and the Omskiy plant fabricated the capsules for landers and the apex cover of the rockets.

These plants were well known for their high level of technological expertise, and we could count on them. Afanasiev's assistance was not only limited to the issuing of decrees on fabrication of the spacecraft parts. He knew that his orders might be handled in different and not always appropriate ways. Afanasiev was concerned with the complexity of situation and at the same time was aware that the failures were unacceptable.

Therefore, for the critical period, he delegated to the Lavochkin design bureau the main control officer of his ministry, I.N. Fedchenko. Fedchenko was well known for his "dog's grasp." Fedchenko took under his control the Urals and Siberian plants as well as the experimental plant in the Lavochkin design bureau. It did not take a long time to see the results of his efforts. Soon, the parts of the spacecraft were delivered to the experimental plant, and the assemblage and testing of the main systems were begun.

The testing operations required that the technological decisions were implemented correctly and that our designers and technicians were highly qualified. The cold dumping and fire testing of the propulsion system were successful. Thermovacuum testing of the duplicate spacecraft ensured that the thermal calculations were made correctly.

We could not wait until the assembly of the spacecraft would be finalized and vibrostatic testing performed. The aircraft, developed earlier in the Lavochkin design bureau were tested differently, and we were not experienced in the vibration testing of big and heavy craft.

At that time, our organization did not possess the vibration exciters with the necessary power. Therefore, the vibration testing of the M-69 spacecraft was performed in a well-equipped test facility of Scientific Production Association (NPO) of Machine Building in the town of Reutovo. At the beginning of testing, all systems performed well. However, in a short time, the vibration of the modules with the scientific instruments, accompanied by an unbelievable noise, started to increase dramatically. It seemed that the modules would fall off. Eventually, the critical part of the test was completed. Nevertheless, the modules did not collapse. Then, the noise evolved into a high-pitch whistle, the spacecraft became almost motionless, and one could not observe any sign of vibration.

In this normal working environment as if on a magic command, the brackets holding the micro-engines for the orientation system started to fall off on the floor one after another like ripe prunes. One glance at the fracture area was enough to know the cause of the destruction. A fatigued metal strip caused the problem. Perhaps the designer had missed out in the college lectures and practical training on this subject. To avoid the destruction of the brackets, it was important to increase the radius of the strip and to refine it.

Further testing operations caused no problems. The head of the materials strength division, Kh.S. Bleikh, was satisfied. His colleagues passed the exam with a grade of A. The vibration testing of the duplicate spacecraft was performed without the lander. At that time, the development of the lander design was discontinued because the weight of the spacecraft construction and onboard instruments exceeded the acceptable limits. In addition, we did not have enough energy and time to continue the labor-intensive balloon testing of the parachute system designed for the lander to descend.

The Earth-based testing operations were for most part completed, and the fabrication of the spacecraft systems would be finished soon. Now, as soon as the fabrication of onboard instruments was completed and they were delivered to the Lavochkin design bureau, the spacecraft would be launched.

3.4 The Onboard Instruments of Spacecraft M-69

The body of the spacecraft is its mechanical design, and the soul of the spacecraft is its instruments and systems.

The brain of the spacecraft is its control system. It directs the spacecraft to the pre-determined position in space and holds it in this position while the engines work. Besides that, the control system measures the thrust impulse, and after the spacecraft achieves a predetermined speed, it executes the command to turn it off. To deliver the spacecraft to the planet, the control system should have a high precision.

The attitude system serves as the eyes of the spacecraft. The spacecraft instruments have to be able to watch the Sun and to be able to find, among the billions of stars, the only one that matters, Canopus.

At the same time, these instruments should not lose sight of Earth, which at the distance of tens of thousands of kilometers appears to be a bright star. The attitude system should be able to maintain the basic reference system with a high degree of accuracy. Based on this reference system, the control system will be able to perform its functions.

The radiotelemetry system performs the function of the tongue and ears of the spacecraft. It receives and transmits to the spacecraft the commands from Earth. At the same time, it receives the information from the spacecraft systems and transmits it back to Earth. In addition, the radio telemetry system measures the radial distance and the speed of the spacecraft. The Earth-based facility allows one to define the position of the spacecraft in space.

The power supply system serves as the blood circulatory system of the spacecraft. It transfers the Sun's rays into the electrical power that supplies all of the spacecraft systems on board.

The propulsion system performs as the legs of the spacecraft. Pushing away the gases discharged from the engine nozzle, the propulsion system provides the spacecraft with the opportunity to move in space.

The fur coat of the spacecraft is its screen-vacuum thermo-insulation system. It protects the spacecraft from the Sun's heat and controls its temperature in extremely cool interplanetary space. The temperature control system keeps the temperature of the spacecraft and its instruments in the predetermined range. If the temperature in the module decreases, the temperature control system, by converting the Sun's heat into the electrical power, will increase it. If onboard instruments are overheated, the temperature control system will discharge the unnecessary heat into interplanetary space.

New significantly advanced, compared to the first-generation spacecraft, multipurpose board systems with improved technical characteristics were developed for the M-69 spacecraft. After acquiring expertise in the design of the control and orientation systems for the Venus 4 spacecraft, our specialists began the development of these systems for the M-69 spacecraft.

The head of the division, S.D. Kulikov (now the main designer and executive director of the Lavochkin design bureau), was ordered to develop the control system. The control system included the gyros that were composed of two free gimbals for measuring the normal velocity, a gyroscope to measure the longitudinal acceleration, sensors to measure the angular velocities, the unit for amplifying the signal, the logic unit, and the operational tools. Gyroscopes were developed in the Scientific Research Institute of the Applied Mechanics, which was headed by Academician V.I Kuznetsov. The system provided the attitude stabilization of the spacecraft in the active periods of interplanetary flight as well as after separating from the booster block D and entering the interplanetary trajectory.

The head of the division, A.S. Demekhin, was in charge of the development of the attitude control system. It was quite different from the attitude control system of the first-generation spacecraft, in which the Sun and star combined sensor, with a field of view of a little more than a hemisphere, was used. The presence of any objects in the hemisphere was forbidden because they could reflect sunlight to the quartz spherical hood of the device. As a result, a star sensor, whose optical tube could be in any position in the spherical belt of ±15 degrees width, can fail. For the same reason, it was not feasible to install a duplicate sensor in the spacecraft.

Apparently, the ability of the spacecraft to function properly depended entirely on the working capacity of one device. Demekhin suggested separating the Sun and the star sensors. Now the tube of the star sensor could move only in a solid angle within ±5 degrees, and protection from the sunlight patches would be simplified. In addition, the number of the forbidden areas where the star sensor cannot perform in the planet's satellite orbit would be reduced, and the opportunity to fabricate the duplicate sensors would be possible. The system used two Sun sensors with a permanent orientation, two Sun sensors with a precise orientation, two star sensors, one Earth sensor, and one Mars sensor. The Sun and star sensors were developed in the Central Design Bureau (TsKB) Geophysics, which belonged to the Ministry of Defense and had extensive experience in the fabrication of optical instruments with a high degree of accuracy.

The specialists from TsKB Geophysics clearly understood that it was extremely difficult to fabricate the variety of optical instruments in less than 1 year. Therefore, they attempted to simplify the design using the decisions checked before. These intentions were not welcomed by Demekhin, who required the accomplishment of the predetermined technical instructions.

The problems of transmitting the control commands from the Sun sensor were discussed and argued for a long time. In accord with the technical assignment, the Sun sensor should have a few zones. While, because of the short-term insertion of the micro-engine, the Sun moved to the edge of the sensor and passed over a few zones, the sensor would generate the signal to relocate the spacecraft to a position where the Sun would return to the center of the sensor. It was expected that, while the Sun moves to the center of the sensor, the signal would be generated only after the Sun moves through several zones. If this design works, the requirements for the attitude system would be reduced.

To simplify the design, the specialists from TsKB Geophysics proposed decreasing the number of zones in the Sun sensor. With that, the signal generated after the insertion of the micro-engine would appear independently of the direction of the Sun's movement. An extended discussion was finished only after Babakin became involved. He convinced the directors from TsKB Geophysics that our requests should be completed. At that time, TsKB Geophysics was overloaded with orders to develop the optical instruments. To facilitate their schedule, the Ministry of Defense transferred the development of the Earth sensor in Kiev to the Central Design Bureau (TsKB) Arsenal.

At that time, the jobs performed under the VPK Resolution were not covered by the contract. After the completion of the job, the contractor should send the bill to the organization that proposed the job. TsKB Geophysics sent us a bill for 300,000 rubles for the development and fabrication of the set of extremely intricate optical instruments. Simultaneously, TsKB Arsenal asked us to pay 3 million rubles for one optical sensor with a more simple design! Naturally, I was outraged with this attitude of robbery.

But our main business manager, F.I. Mitelman, made the philosophical remark: "Everything is clear. The Geophysics Bureau has plenty of jobs and put their real expenses in the bill for each job. However, the number of jobs in TsKB Arsenal was limited and their managers put in our bills all their spending costs. We will not be able to prove anything. Calm down and send them the check." Such was the ugly reality of Soviet economics.

The major elements of the attitude system were the micro-engines that worked on high-pressure nitrogen. The requirements applied to them were harsh. It was expected that after a half a million insertions, the micro-engines would maintain their characteristics and would produce the thrust impulses within specified limits of the leading and trailing edges of the front. The study showed that the characteristics of the micro-engines could be maintained if a metallic gasket was used in the valve. The development of the micro-engines was assigned to the Central Scientific Research Institute of the Fuel Automatics (TsNII TA), which belonged to the Ministry of the Automobile Industry and was located in Leningrad.

The enthusiastic professional and head of the department in TsNII TA, A.V Presnyakov, was in charge of this project. The specialists from TsNII TA were not afraid that the micro-engines were expected to be inserted many thousands of times. In automobile designs, even millions of engine insertions are a common occurrence. Soon the technological instructions were developed, and the set of the micro-engines was fabricated. The test experiments showed that the micro-engines were in compliance with our requirements.

Thereafter, TsNII TA transferred the technological documentation for the fabrication of the micro-engines to the experimental plant of the Lavochkin design bureau. The first set of micro-engines fabricated and assembled in the experimental plant was in compliance with all requirements, except one major point. After a few thousand insertions, the hermetical seal of the double valves were destroyed. As a result, during a long space flight, the nitrogen retained in the attitude system could be vented.

The specialists from TsNII TA believed that the thin layer of grease over the valve plate could not be the reason of the flaw that was discovered, and they delegated their technician to our organization. Upon arrival, the technician commented on the good quality of the valve surface, degreased the plate surface of the valve, pulled out of his pocket a wooden beech board, covered it with tracing paper, and rubbed with the plate against it. After this procedure, an almost undetectable grease layer remained on the plate. Micro-engines assembled according to this technological process worked perfectly, proving once again the correctness of the Russian proverb "A good master does good work."

The radiotelemetry system was developed in the Central Scientific Research Institute of the Space Instruments Development (NII KP) under the supervision of M.S. Ryasanskiy, the main designer and correspondent member of the Academy of Sciences of the Soviet Union. The radiotelemetry system consisted of:

1. The transponder-receiver, which worked in the frequency band 790-940 MHz. It was designed to receive radio commands, to measure the radial distance and velocity, and to transfer the telemetry data. The onboard transponder operated at 100 watts of power, and data were transmitted at a rate of 128 bits/sec.

2. The impulse transmitter, which worked at a frequency of 6 GHz. It was designed to transmit images of the Martian surface to Earth. The transmitter operated at a power of 25 kilowatts, and data were transmitted at a rate of 6,000 bits/sec.

3. The telemetry system, designed with 500 channels to provide the data from the onboard systems.

4. The antenna unit, which included three low-gain antennas in the decimeter band, a high-gain antenna with a diameter of 2.8 meters for the decimeter, and centimeter bands and other parts.

The antenna system was developed and fabricated in the Lavochkin design bureau according to technical documentation of NII KP. The total weight of the radiotelemetry system, including the antenna unit, was 212 kilograms. In addition to the radiotelemetry system, NII KP developed a camera (FTU) for acquiring the images of the Martian surface. This intricate and clever device consisted of:

1. A film bobbin, whose sensitivity was artificially reduced to avoid exposure to radiation
2. A unit designed to restore the film sensitivity
3. A unit for film processing
4. A unit for exposure
5. A data encoder

It is worth noting that in spite of the complex design, the FTU worked perfectly. Each device could store 160 images. Each image was made with 1,024 x 1,024 pixels.

The power system was developed in the Lavochkin design bureau under the supervision of N.F. Myasnikov, who was the head of the department. The power was provided by solar panels with an area of 7 square meters, which was designed in the Scientific Research Institute of Current under the supervision of N.S. Lidorenko, the director, main designer, and corresponding member of the Academy of Sciences of the Soviet Union. The system included a hermetically sealed cadmium-nickel battery with a capacity of 110 amp hours. This battery was designed by the Scientific Research Institute of the Battery in Leningrad.

The development of the Project M-69 systems was based on recent technical achievements. However, the data processing system designed for the processing of scientific information was more advanced. This system was developed at NPO under the supervision of G.Ya. Guskov, the director, main designer, and corresponding member of the Academy of Sciences of the Soviet Union. Guskov was in charge of a team of young and very energetic engineers.

The Guskov NPO was located in the town of Zelenograd, which at that time was the center of a rapidly developing Soviet electronic industry. By using a special technology, which in the Institutes of Microelectronics was not yet fully developed, the NPO provided a significant breakthrough in science and technology. In addition to the data acquisition, the system was able to program the scientific instruments and to process and compress the data transmitted from the instruments. The system weighed only 11 kilograms.

Ryasanskiy did not want to have a competitor who had access to advanced technology and therefore could push him out of important developments in space exploration. He asked Babakin, with whom he had close personal relations, not to involve Guskov in the development of interplanetary spacecraft. For a while thereafter, everything was calm. But as the say, "You cannot hide a needle in the haystack."

At one of the meetings, the representative of the Institute of Space Research mentioned that Guskov was developing a system for processing scientific data. Ryasanskiy started to worry and after the meeting approached Babakin and said, "Georgy, you promised me that Guskov would not be involved in this job." Babakin responded cunningly, "Don't worry, Misha; he is only involved with the development of a small block."

Although official ideology always rejected tough competition between the institutes and the design bureau, in technical circles it always existed. Being a young specialist, I designed a high-quality air regulator for pressurization of the fuel tanks, which was installed in the LA-250A airplane. G.I. Voronin, who was the main designer of OKB-124, which developed parts for airplanes, learned about my device. To be able to use advanced technological solutions, he delegated to S.A. Lavochkin, the head of the department, and instructed him to learn about the design of the air regulator.

The leader of my department, N.N Gorshkov, gave me a letter from G.I. Voronin, endorsed by S.A. Lavochkin, and said, "Semen Alekseevich asked you to give him information but to do it in such a way that he would not be able to learn anything." It was my first vivid example of direct competition between two organizations of the same ministry.

3.5   The Beginning and End of Project M-69

At the end of the third quarter of 1968, we were quite behind schedule in the development of systems and parts of the spacecraft and their fabrication. This trend had to be changed immediately, because otherwise the launch of the spacecraft to Mars would inevitably be postponed until 1971. The completion of this task was important from both scientific and political points of view.

As it was commonly practiced at that time, employee meetings were held to boost morale and to accomplish the task at hand. If the decisions of the employee meetings were not fulfilled, it would probably not have any serious consequences. However, if the resolution of the Communist Party meeting was not fulfilled, in the best-case scenario, one would be penalized by the leaders of the Communist Party. In the worst-case scenario, the head of organization might be eliminated from the Party and moved to a lower level position.

After the Communist Party meetings, the managers of all levels focused their efforts on the completion Project M-69. It should be mentioned that the employee meetings were useful and made people work hard. After these meetings, most of the engineers, technicians, and workers were anxious to complete the job and to launch the spacecraft on schedule. The specialists from design bureaus were working round the clock, sleeping only for a few hours a day at their workplace on folding beds. The local cafeteria was ordered to stay open 24 hours a day and provide free meals to the employees. In spite of the overwhelming work effort, no proper compensation was provided. The people's enthusiasm was considered to be the major driving force.

The Communist Party Central Committee was rather concerned with falling behind the work schedule and the potential failure to achieve the political goals. It was decided to monitor the progress of work on Project M-69 on the weekly basis. Heads of organizations who had not accomplished their goals on time were called on for explanations. The blacklist of underperformers was compiled before the meeting. We understood the pressures imposed on the heads of those organizations. If we were certain that they would completely fulfill their orders for onboard equipment, we would not include them on the blacklist.

This strategy corrected the situation. The first sets of the onboard instruments were delivered to the Lavochkin design bureau and to the Institute of Space Research. The control and attitude systems started to be tested, and the assembly of the first spacecraft began. The equipment for the second spacecraft was delivered with approximately a 1-month delay. Finally, after testing in the Institute of Space Research, the module with the scientific instruments and the data processing system was delivered. E.M. Vasiliev, who was the head of the department in the Institute of Space Research, handed it over to the Lavochkin design bureau on December 31, 1968, at 11:00 p.m.

In the middle of the January 1969, the testing of the first spacecraft and the assembly of the second spacecraft were completed. The time that remained until the launching date was limited. That was why the decision was made that the plant and ground stages of the testing operations of the power system would be combined for the second spacecraft. Both spacecraft were sent to the Baikonur launch complex.

However, Mars, which did not want to open its secrets to humanity, tested the team that prepared the spacecraft for launching. At the end of February, during the launch operations, the powerful N-1 rocket blew up. The strong explosion blew out the windows in all the hotels. The weather was extremely cold, the temperature outside was –30 degrees Celsius, and the central heating system immediately became frozen. Windows were installed, but it was not possible to replace the damaged pipes and radiators of the central heating system. The electrical heaters that were provided for each hotel room could keep the temperature only a little above zero. Nevertheless, people did not give up and continued to work to prepare the spacecraft for launch. It was a real challenge!

On March 27, 1969, the first spacecraft was launched. The loud speakers announced, "The flight is proceeding normally. The pitching, yawing, and rotation are within the standard limits. The first stage is separated, the apex cover is separated, the second stage is separated, the flight is normal."

After a minute, the speakers fell silent. Thereafter, we heard, "No signal." After the explosion of the third stage of the rocket, the flight was finished. The remains of the spacecraft had fallen in the Altai mountains.

In April 2, 1969, the second spacecraft was launched. After the rocket lifted off, a steam of black smoke appeared in the right engine. A few seconds passed, and an explosion occurred. The rocket was transformed into a dense, luminous bright mass of fire. Everything was finished.

4.1   The Optional Routes to Mars

The year 1971 was one of conjunction, when Earth and Mars were at a minimum distance from each other. That happens once in 15 to 17 years and is a very favorable time for Mars observation and for interplanetary flights as well. During these years, minimum power is required to deliver the spacecraft to Mars, and, consequently, a heavier spacecraft can be launched on an interplanetary trajectory.

In 1969, if the launch of the M-69 spacecraft had been successful, we would have acquired detailed data on the Martian ephemeris and on atmosphere pressure near the Martian surface. We intended to use these data in 1971 to make a soft landing on Mars.

The unsuccessful launches of the M-69 spacecraft put us in difficult position. We could postpone the program until 1971 and undertake Mars exploration from a satellite orbit and with an entry probe. In this case, we would repeat the program initially planned for the M-69 spacecraft. Thereafter, in 1973, we could launch a spacecraft, consisting of an orbiter and a lander, to Mars.

It was proposed that after a soft landing, the lander would make a scientific exploration of the Martian surface. After careful consideration, a delay of the program was abandoned. This decision was made because, in 1973, the power required for launching the spacecraft had to be increased, and as a result, one rocket would not be able to launch both an orbiter and a lander to Mars. Launching an orbiter and a lander separately would be very expensive and intricate and would require four launches.

Eventually, the appropriate decision was made. The main point was that three spacecraft should be launched to Mars in 1971. A spacecraft with a very large supply of fuel and the maximum amount of scientific instruments on board should be launched first. This particular spacecraft would reach Mars significantly earlier than the next two and would be placed in a Mars satellite orbit. Also, it would serve as a radar beacon for the second and third spacecraft.

The measurements made by the orbiter on its outbound trajectory would allow for the exact determination of the position of the planet. In addition, it would be possible to eliminate the error caused by the lack of a precise Martian ephemeris and to precisely define the entrance corridor in which the lander enters the Martian atmosphere.

This proposal was accepted because it would define the appropriate angle at which the lander would enter the Martian atmosphere and would provide the opportunity to launch the first satellite to Mars before the United States would launch their Mariner spacecraft.

At the end of May 1969, M.V. Keldysh, the President of Academy of Sciences of the Soviet Union, led a meeting at which our proposal to launch three Martian spacecraft was discussed. As usual, the meeting was not conducted in a building of the Presidium of Academy of Sciences of the Soviet Union. Keldysh held it in his small and cozy office in the Institute of Applied Mathematics located in Miusskaya Square. The number of participants was limited and included G.A. Tyulin, the first deputy-Minister of the Ministry of General Machine Building, E.N. Bogomolov, representative of the Ministry of General Machine Building, Academician A.P. Vinogradov, Yu.A. Surkov, the representative of the Institute of Geochemistry, V.I. Moroz and E.M. Vasiliev, representatives of the Institute of Space Research, M.S. Ryasanskiy and Yu.F. Makarov, representatives of NII KP, M.Ya. Marov, representative of the Institute of Applied Mathematics, and G.N. Babakin and V.G. Perminov, representatives of the Lavochkin design bureau.

Keldysh was the main driving force of all space programs in the Soviet Union. He participated in each meeting vital to the space program. Keldysh cherished Babakin, who for a short period of time became very successful in the exploration of the Moon and Venus. Keldysh believed in his talent and supported his daring projects. After careful and detailed consideration, our project was approved. At the end of meeting, Ryasanskiy addressed the audience. He suggested that efforts of specialists be focused and that costly resources be avoided. In his view, to process the scientific data, the telemetry unit in the radiotelemetry system should replace the telemetry system developed by Guskov for the M-69 spacecraft.

At first glance, this idea seemed to be reasonable, and the participants of the meeting approved it. That was because no one except Babakin, Vasiliev, and myself was aware of the characteristics and operational functions of this system. Only Babakin could defend the Guskov system, but he kept silent. Technical progress was sacrificed because of his friendship with Ryasanskiy.

4.2 The Decision Is Made: The Search Continues

If the first M-71S spacecraft could be launched ahead of schedule, about 800 kilograms of fuel was needed for its tanks. The current spherical tank designed for the M-69 spacecraft needed to be upgraded because the increased amount of fuel could not be placed there. There were two solutions regarding how to increase the volume of the fuel tank:

1. Increase the diameter of the tank or
2. Insert two circular plugs in the current tank so that it will acquire the shape of an egg

Neither solution was very attractive. The fuel tank was a major part of the spacecraft. Modules with instruments and other parts were attached to it. If the design of the tank was changed, the design of the spacecraft should be changed as well. In addition, to be sure that the changes in the design are made correctly, Earth-based testing operations must be conducted.

However, considering the large amount of future modifications, it was unreasonable to maintain the design of the hermetically sealed cylindrical modules with the instruments developed for spacecraft M-69. Naturally, we had a lot of trouble with them during the electrical-radio-technical testing operations. If any unit stopped performing, the whole onboard cable network had to be disconnected from the hermetically sealed sockets and the frame with instruments had to be removed from the module. If the frame had been located away from the spacecraft, the access to the instruments would have been excellent. However, after testing or replacing the device, the frame with the instruments had to be placed back in the hermetically sealed module. The cable network should be connected to the hermetically sealed sockets, and all autonomous testing operations should be repeated. These procedures led to a great waste of time and to the appearance of hidden flaws.

After considering all the pros and cons, we decided to modify the configuration design of the spacecraft again. Several designers were challenged to find the best solutions. B.N. Martynov, the head of the design division, and I very carefully analyzed the proposals and, if negative points were revealed, rejected them. Eventually, we approved the configuration design of the spacecraft proposed by a young designer, V.A. Asyushkin (now he is the main designer of the booster rocket Fregat).

The configuration of the spacecraft was in complete accord with the major instructions. To achieve the different tasks, the spacecraft could be easily modified without changes in the main design. The propulsion unit was the major element of the spacecraft. It was composed of a separate module with a cylindrical fuel tank. The fuel tank was divided into two tanks consisting of fuel and oxidizer. With a gimbal, the engine was mounted to the lower surface of the tank. The solar panels, the high-gain antenna, and the radiators of the temperature control system were attached to the cylindrical part of the tank. The lander was attached to the upper surface of the tank. The instrument module with a separable lower cover was attached to the lower part of the tank.

This design allowed unlimited access to the onboard instruments, located in the module. Because of the increased amount of fuel, the spacecraft configuration would not be modified; only the length of fuel tank had to be changed. For a long time this successful design was used not only for Mars exploration but also for the exploration of Venus and Halley's comet and for astrophysical investigations from an Earth satellite orbit.

Work was proceeding at a slow pace on increasing the accuracy of the control system and decreasing the second-order error terms in the angle at which the lander would enter the Martian atmosphere. The department headed by S.D. Kulikov was again focused on the development of the spacecraft Prognos. Simultaneously, the development of the control system for the Martian spacecraft had been transferred to the department of A.S. Demikhin, where specialists to carry out the whole spectrum of tasks were in short supply.

Babakin decided to transfer the development of the control system to the NPO of the Automatics and the Instruments Development (NPO AP). In the Soviet Union, NPO AP was the largest company that specialized in the development of control systems for spacecraft. The head of NPO AP was N.A. Pilyugin, an Academician and the main designer. He was a member of the legendary team of five main designers who used to work with Korolev. Besides Korolev and Pilyugin, the team included Academicians V.I. Kuznetsov and V.P. Barmin and corresponding member M.S. Ryasanskiy.

After an evaluation of our technical assignment, the NPO AP specialists suggested their own digital version of the control system. The control system that had been developed under the supervision of Pilyugin for the last stage of the N-1 rocket was suggested as the prototype. Unanimously, this system was rejected by our specialists. It was heavier and required more power than the control system of the M-69 spacecraft.

Pilyugin agreed to develop the new system with suitable power and weight parameters. But he was able to finish this job only in 1973. For us, that was absolutely unacceptable. Only Babakin and Pilyugin participated in the further meetings at which the development of the control system for Project M-71 was considered. We were listening to each other's arguments, evaluating them, meeting again, and just could not reach an agreement. At one of the meetings, Pilyugin, usually calm and reserved, quite seriously suggested that Babakin fire Demekhin, who was the main opponent of the new system.

Eventually, our head managers got tired of the ongoing battles. They secluded themselves from others for half an hour and came out with a joint decision. The control system suggested by Pilyugin would be installed in the M-71 spacecraft. To console us, Pilyugin agreed to fabricate the automatic blocks of the spacecraft at his experimental plant according to our technical instructions (for NPO AP, it was an unheard-of compliance).

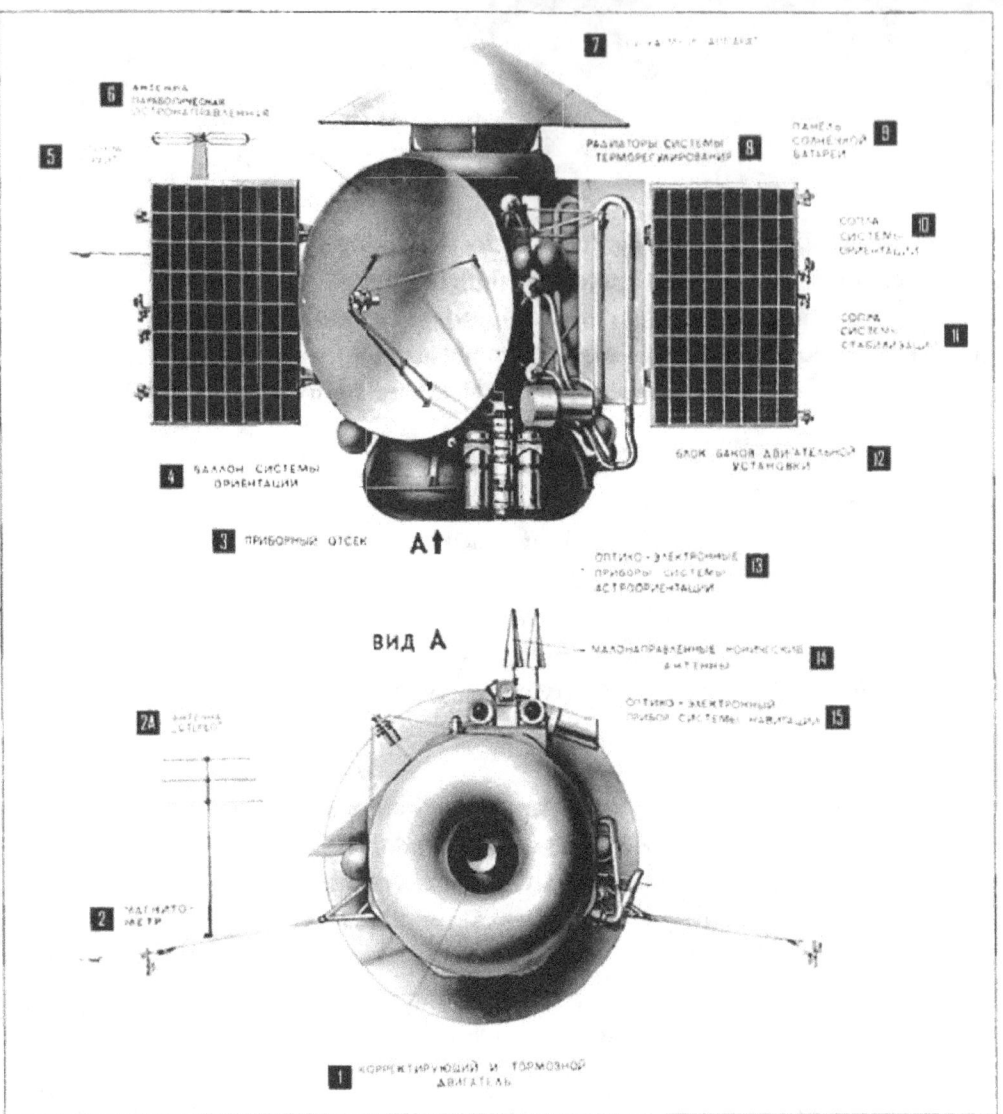

We growled at Babakin as soldiers would growl at their general who lost the battle. Only 1 year later, we realized how clever our leaders were. This control system provided the opportunity of a soft landing and made it feasible to use the M-71 spacecraft in future projects. The control system weighed 167 kilograms, consumed 800 watts, and provided the necessary accuracy for maneuvering of the spacecraft in its trajectory.

The spacecraft weight was decreased because of the new control system. To make the loss in weight minimal, we removed the automatic control system from the booster block D. Now the booster was controlled by the spacecraft's control system (Figure 6).

Figure 6.
The Mars 3 Spacecraft —
(1) correction braking engine, (2) magnetometer, (2A) stereo antenna, (3) module with the instruments, (4) orientation system, (5) stereo antenna, (6) high-gain parabolic antenna, (7) lander, (8) radiators of the temperature control system, (9) solar panel, (10) nozzles of the attitude system, (11) nozzles of the stabilization system, (12) unit of tanks of the propulsion system, (13) optical-electronic devices of the astro navigation system, (14) low-gain conical antennas, and (15) optical-electronic device of the navigation system

4.3 The Lander: The Search for the Best Decision

Scientifically and technically, the development of the lander design was a very complex problem. To solve it, one should search for the best decision. First of all, it had to be decided how the lander would descend in the Martian atmosphere. Which type of descent should it be? Gliding or ballistic? If the gliding option would be chosen, the lander's gravity center with respect to the axis of symmetry of the brake cone should be changed. Under the influence of aerodynamic forces, the lander would be stabilized at some angle of attack to the approaching airflow. As a result, a gravitational force (G-force) would be generated, and the length of the braking path and altitude, at which the parachute would be deployed, would be increased.

In 1975, this landing design was used by American specialists in the Viking Project. Because of the lack of the detailed data on the Martian atmosphere, we were unable to use the gliding descent.

Therefore, for Project M-71, a ballistic descent was chosen. The trajectory would be entirely defined by the initial conditions under which the lander would enter the Martian atmosphere and by the ballistic coefficient that is directly proportional to the area of the break cone. If other conditions are equal, the larger the ballistic coefficient, the earlier the parachute system would be open and the higher the altitude.

Aerodynamic modeling of the conditions in which the lander would descend indicated descent with a break cone with a vertex angle of 120 degrees and with a diameter of 3.2 meters. The outer diameter was restricted by the diameter of the apex cover of the Proton rocket. The angle of the cone opening had been chosen to yield the maximum coefficient for the front resistance and to preserve the stabilization of the lander during its descent in the Martian atmosphere. After ballistic braking, and at a descent velocity of Mach 3.5, the parachute would be deployed.

The parachute system was designed in the Scientific Research Institute of the Parachute Landing Facilities (NII PDS) under the supervision of N.A. Lobanov, who was the director and the main designer at the institute. NII PDC had a high level of expertise in the development of the multipurpose parachute systems, including a parachute system used above the velocity of sound.

But what was required for Project M-71 went beyond all the previous systems. Never in the aerospace history had any parachute system been deployed at such low atmosphere pressure and at such high flight velocity. Theoretical calculations and tests of different parachute systems in aerodynamic pipes had been performed. Based on this study, the NII PDS specialists suggested that Project M-71 use a parachute system consisting of an auxiliary parachute with an area of 13 square meters and a main parachute with an area of 140 square meters.

The testing operations were an important stage in the design of the parachute system and were performed to prove its proper deployment during flight. The testing procedures for the parachute system carrying the probe had already been developed during Project M-69. At that time, we decided to use a small balloon to lift the model probe with gunpowder accelerators at an altitude of 32 kilometers above Earth's surface. Using gunpowder accelerators, after separation from the balloon, the probe sped up until a predetermined velocity at which the parachute system would open.

While developing the parachute system for the Viking lander, American specialists independently made the same choice of lander testing operations. That did not console us. We recalled that the main reason for eliminating the probe device from Project M-69 was the necessity to perform intricate experiments, which simulated the conditions of the deployment of the parachute system with big balloons. Fearing a repetition of the same events in Project M-71, the decision was made to study other versions of parachute systems.

The mutual efforts of the specialists from the Lavochkin design bureau and NII PDC bore fruit. It was proven that reliable information about the performance of the parachute system could be obtained on test models of the lander with a scale factor of 1:5. The test model was lifted to an altitude of 130 kilometers by the gunpowder meteorological rocket M-100B. In free fall, the model achieved the velocity at which the parachute system was activated.

An informal team of young engineers with a high level of theoretical expertise made an important contribution to solving the aerodynamic problems of project M-71. The team consisted of Yu.N. Koptev (now the general director of the Russian Space Agency), N.A. Morozov and V.V. Kusnetsov. I recall that members of this team once worked in different departments of the Design Bureau. Apparently they got satisfaction by solving problems and that united them in their mutual work.

A soft landing is the most complex stage of any flight. Being aware of this, passengers nervously await for their airplane to land on a runway equipped with modern technological facilities. With this in mind, one can easily understand the problems faced by the lander designers.

Unlike an airport, there is no flat runway or any kind of landing area. Instead, there are areas covered with sand and possibly with stones and perhaps having steep slopes. Also, one must consider wind of unknown direction and velocity. In addition, there is no pilot.

Nevertheless, to measure the lander velocity and its drift direction, it would be useful to have a velocity meter working on the Doppler principle. The meter would direct the commands to the engine, which controls the soft landing. Then the drift and decrease in vertical speed could be corrected. The only thing left to do was to offer a prayer that the landing area would be flat and contain no stones.

Unfortunately, because of the insufficient weight of the lander, a descent with instruments working on Doppler principles was not feasible. Therefore, we decided to utilize another descent design. When the lander approaches the Martian surface, a radar-altimeter, at the appropriate altitude and velocity of descent, will direct commands for the insertion and deactivation of the soft landing engine. The major component of the lander, the automatic Martian station, would fall free to the surface from a low altitude. The lander has to be durable enough to sustain the impact of this free fall landing.

We attempted to use rubber air bags, which were already used for shock absorption of the automatic lander on the Luna 9 mission. However, we were unable to protect air bags from the stream of hot gases coming from the gunpowder engine. Other methods of shock absorption for the Martian station, including one using a special nose cone spear, were not considered because they could not make the automatic Martian station secure in all simulated situations.

Babakin approved my idea to use foam plastic for the protection of the station during landing. I suggested that the automatic Martian station should be covered by foam plastic on each side. Design experiments indicated that the foam plastic

Figure 7.

The M-71 Lander—(1) instruments of
the automatic control system,
(2) nitrogen container, (3) braking
cone, (4) radar-altimeter antenna,
(5) parachute-instrument module,
(6) antenna for communication with
the Martian satellite, (7) engine that
activates the auxiliary parachute,
(8) joint frame, (9) engine that
initiates the landing, (10) main parachute, (11) engine that controls the
pitch and yaw of the lander,
(12) container of the scientific instruments, (13) automatic Martian
station, (14) instruments of the
automatic control system,
(15) engine for the parachute withdrawal, (16) propulsion system for
soft landing (17) nitrogen container
for the control system

should cover the lower part of station, which faced the main impact during landing. In the upper part of the station, the foam plastic blanketed a special aeroshell cover, which protected the systems and scientific instruments from damage. Therefore, during landing on the Martian surface, the lander was pro-tected from impacts and damage from every side.

4.4   The Lander Design and Its Mission

The lander that had been designed was a compact design and based on a joint frame (Figure 7). The gunpowder engine, which would transfer the lander from the flyby trajectory to the Martian encounter, was attached to the joint frame. The instruments and parts of the automatic control system as well as the containers with high-pressure nitrogen would be attached to the joint frame. This equipment would provide the essential atti-tude control for the lander during its autonomous flight.

In these plans, the joint frame was attached to the brak-ing cone by four bars. The gas micro-engines, which would provide the attitude control of the lander during its autono-mous flight near the planet, were installed on the bars. Four gun-powder micro-engines were attached to the outer part of the braking cone. Two micro-engines were used for controlling

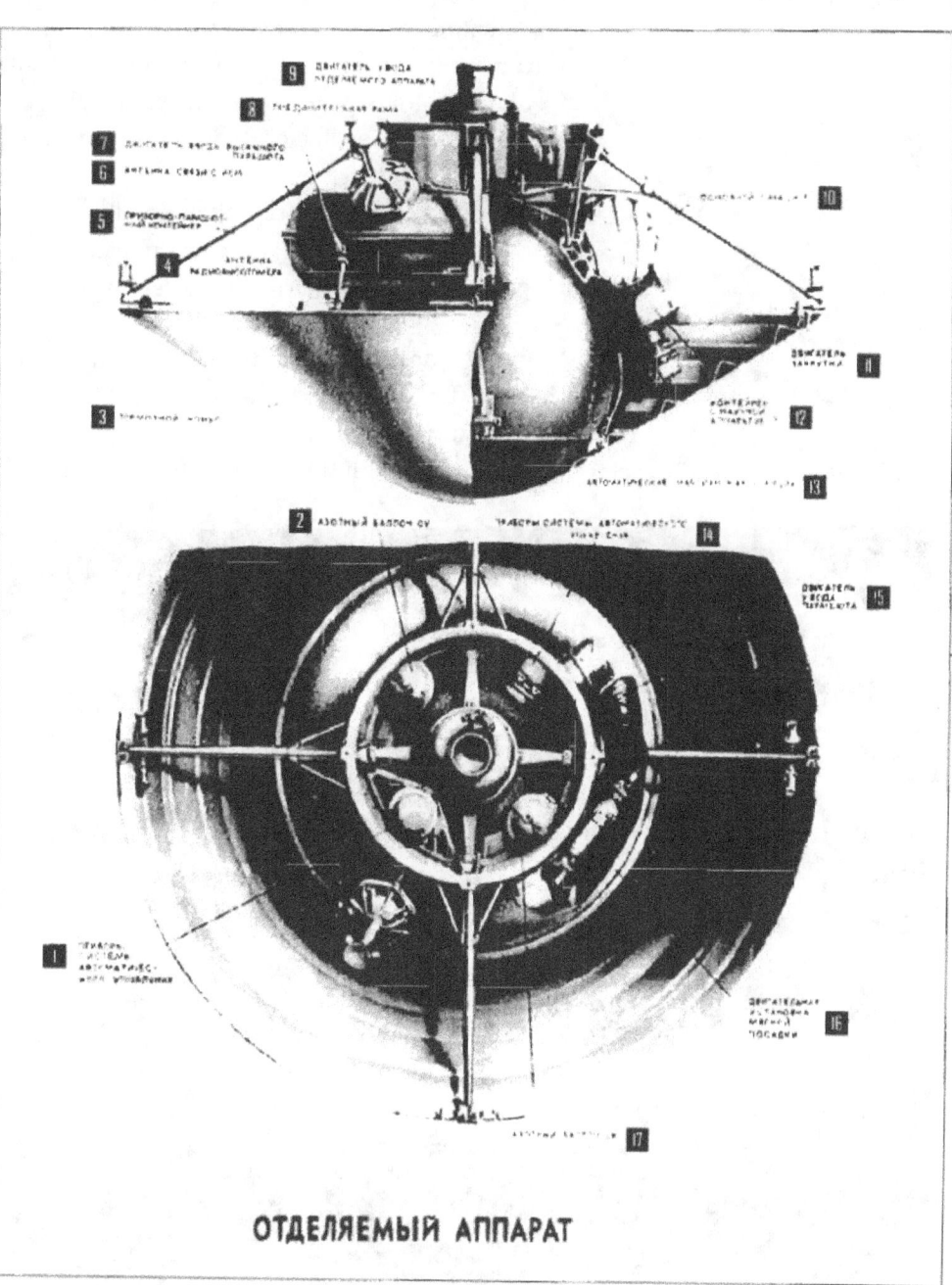

ОТДЕЛЯЕМЫЙ АППАРАТ

pitch and two micro-engines were used to control yaw of the lander with respect to its longitudinal axis.

It was proposed that the automatic Martian station would be located in the braking cone. The parachute-instrument module was attached to the station by metallic ribbons. It included the auxiliary and main parachutes, the gunpowder engines that provided the soft landing, and the withdrawal of the parachute. The engine that controlled the auxiliary parachute had four lateral nozzles and was attached from the outside.

The scientific instruments would be placed in the lower part of the parachute-instrument module in a separate cylindrical container. They could be used to study the Martian atmosphere during the parachute part of the descent. In a toroidal (donut-shaped) area, the parachute-instrument module was divided into two parts.

In the upper cover of the container parallel to the divider, an expanded cumulative cartridge was installed. It was used to instantaneously cut the container cover when the parachute system started to be deployed.

All gunpowder engines were developed and fabricated in the Scientific Research Chemical Technological Institute under the supervision of Academician B.P. Zhukov, who was the director and the main designer at the institute. We had good professional relations with this institute from the time Lavochkin was alive and were very often impressed by the courage and calm of people working at such a dangerous production facility.

Once, having arrived at the institute, I found myself as if in a theater. In front of me was a production facility line with working equipment. The wall of the production facility was missing. It was lying nearby on the ground. "Are you trying to remodel the plant while continuing fabrication?," I asked an acquaintance of mine. "No," he said calmly. "Yesterday, the gunpowder cartridge explod-

Figure 8.
The Lander Ready for Docking With the Spacecraft

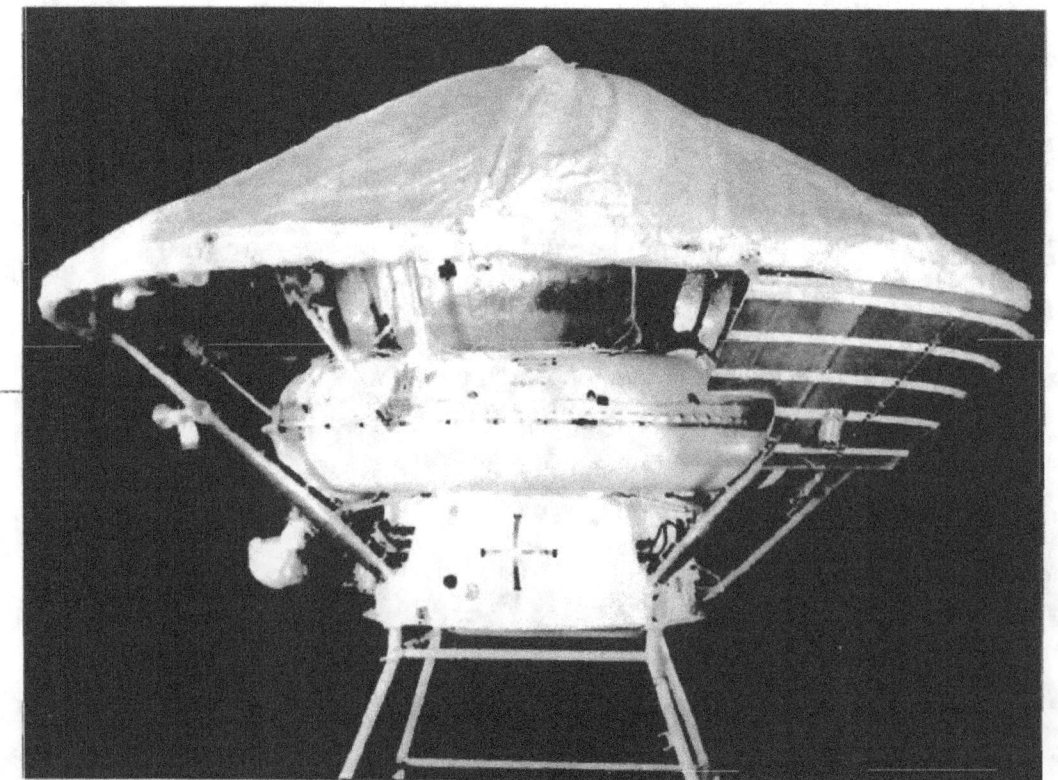

ed and the wall fell down. This wall is specifically designed to collapse as a whole upon impact, so the rest of the production facility would remain intact. Today, they'll put it in place and everything will be in order."

Figure 8 shows the lander prepared for docking with the spacecraft. The surface of the braking cone is covered by the screen for vacuum thermal insulation. On the right part of Figure 8, one can see the radiator for the temperature control system, which provides the proper temperature environment for the lander and the control system in the interplanetary trajectory.

Figure 9 shows a longitudinal section of the automatic Martian station. A layer of foam plastic (2, 19) protects the lander from every side. In the lower part of the station, the foam plastic is 200 millimeters thick. In the upper part of the station, one can see the two-layered aeroshell cover supported by the transverse frames (18). The expanded cartridge (20) is installed in the lower part of the aeroshell cover. When the power system is initiated, the cartridge instantaneously separates the aeroshell cover from the station. Compressed air retained in the circular container (24) is discharged into a displacing bag (17), which ejects the aeroshell cover.

The inner thermal insulation (13) is designed to maintain the appropriate temperature inside the station during the severe Martian nights. After the pyrotechnic lock (10) is broken, the pyrocylinders (15) are activated, opening the four petals (16) and placing the station in a vertical position. The gamma ray spectrometer is installed on one of the petals. The x-ray spectrometer is installed on another petal. Both of them are used to study the composition of the Martian soil. Pyrotechnic devices (11) move the temperature and wind velocity recorders outwards to protect their readings from the impact of the station. Simultaneously with the opening of the petals, the elastic antennas (1) of the radio system are switched into use (Figure 10). The x-ray spectrometer and the instrument (PROP-M) designed for soil penetration are placed on the Martian surface. The PROP-M instrument has a cable 15 meters in length. During its maneuvering, mechanical properties of the Martian soil are measured using the press tool. These data would be used in the future to develop the Martian rover.

The landing program would be started at the distance of 46,000 kilometers from Mars. Using pyrotechnic devices, the lander (Figure 11) and orbiter would separate. In 900 seconds, when the lander is at a safe distance from the orbiter, the gunpowder engine is started. As a result, the lander acquires a velocity of 120 meters per second and is transferred from the flyby trajectory to a Martian encounter trajectory.

In another 150 seconds, in accord with the commands from the automatic control system, the lander would turn to such a position that the direction of its longitudinal axis would coincide with the velocity vector of the moving air stream during the lander's entry into the Martian atmosphere. In this position, the lander, using the gunpowdered engines, turns around its longitudinal axis. The joint frame with the

Figure 9.

Automatic Martian Station of Mars 3—(1) radar-altimeters of the control system, (2) shock absorber of the lower part of the station, (3) telemetric units, (4) automatic radio system, (5) antennas of the radio system, (6) radio system, (7) blocks of the radio system, (8) modules with the scientific instruments (9) telephotometers, (10) lock of petals to place the station in a vertical position, (11) devices to move the scientific instruments outwards, (12) sensors of the scientific instruments, (13) thermo insulation system, (14) screen-vacuum thermo insulation system of the upper part of the station, (15) pyrocylinders to place the station in a vertical position, (16) petals, (17) displacing bag, (18) aeroshell cover, (19) shock absorber of the aeroshell cover, (20) expanded cumulative cartridge for the separation of the aeroshell cover, (21) automatic control system, (22) power system, and (23) receiver of the atmospheric pressure

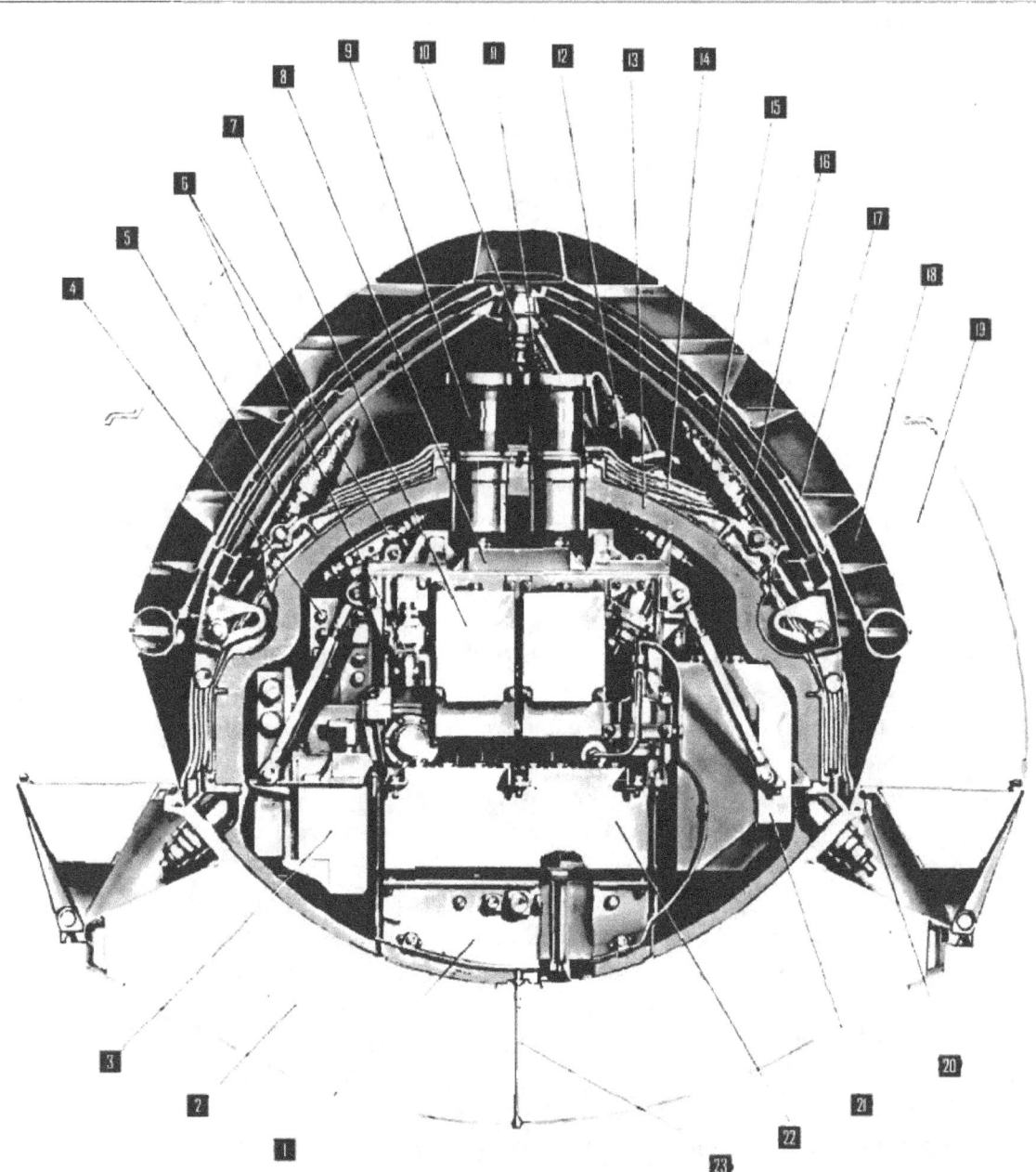

автоматическая станция „М-71„

1-ВЫСОТОМЕРЫ СИСТЕМЫ УПРАВЛЕНИЯ ; 2-АММОРТИЗАЦИЯ НИЖНЕЙ ЧАСТИ КОРПУСА ; 3-БЛОКИ ТЕЛЕМЕТРИИ ; 4-АВТОМАТИКА РАДИОКОМПЛЕКСА ; 5-АНТЕННЫ РАДИОКОМПЛ. ; 6-АФУ РАДИОКОМПЛ ; 7-БЛОКИ РАДИОКОМПЛ ; 8-БЛОКИ НАУЧН.АППАРАТ ; 9-ТЕЛЕФОТОМЕТРЫ ; 10-ЗАМОК ЛЕПЕСТКОВ ВЕРТИКАЛИЗАЦИИ ; 11-МЕХАНИЗМЫ ВЫНОСА НАУЧНОЙ АППАРАТ ; 12-ДАТЧИКИ НАУЧН.АППАРАТ ; 13-ТЕПЛОИЗОЛЯЦ ; 14-ЭВТИ ВЕРХ.ЧАСТИ КОРП ; 15-ПИРОЦИЛИНДРЫ СИСТ ВЕРТИКАЛИЗ ; 16-ЛЕПЕСТКИ СИСТ ВЕРТИК. ; 17-ВЫТЕСН.МЕШОК ; 18-ЗАЩИТН.КОЖУХ ; 19-АММОРТИЗ.ЗАЩИТ.КОЖУХА ; 20-УКЗ ОТДЕЛЕН.ЗАЩИТН.КОЖУХА ; 21-АВТОМАТ.СИСТ УПРАВЛ ; 22-БЛОК ПИТАНИЯ ; 23-ПРИЕМНИК ВОЗДУШН.ДАВЛЕНИЯ

gunpowder engine and the automatic control system, which provided the lander stabilization after its separation from the orbiter, is released. Stabilized by rotation, the lander continues its flight until its encounter with Mars and enters the Martian atmosphere with the speed of 5,800 m/sec. In the Martian atmosphere, the lander would be affected by G-forces generated by aerodynamic braking. In the initial stage of the descent, the G-forces increase. Thereafter, during the lander's speed reduction, the G-forces decrease. Gravitational forces change with the change in the speed of the lander's descent. If the G-force is increased to 2 units, the onboard automatic control system would direct commands to the gunpowdered engines, which would discontinue the lander's rotation. Now the control of the pitch and yaw is not required because during the descent in the atmosphere, the lander would be stabilized by the moving air stream. After about another 100 seconds, the recorder of the relative accelerations (DOU), which controls the ratio of the G-forces at the ascending and descending parts of the curve, would direct a command T1 to start the engine and to activate the auxiliary parachute (Figure 12). In addition, a program timing device (PVM) would be initiated. That happens when the lander speed drops below Mach 3.5. On the PVM command, in 2.1 seconds, the expanded cumulative cartridge would be ignited and would cut the parachute-instrument module. The main para-

chute with the folding shroud lines would be released. In 10 seconds, the main parachute with the withdrawn shroud lines would start to be deployed. The PVM directs a command at T4 and T6 to separate the cone and to turn on the high-altitude radar-altimeter (RVBV). The parachute provides a speed of the descent of not more than 65 meters per second.

Depending on the descent velocity, in an altitude range of 16 to 30 meters, the high-altitude radar-altimeter issues a command; the onboard radio system is disconnected, the engine that controls the soft landing is started, and a second PVM program is initiated. When the lander's vertical speed of descent drops to 6.5 meters per second, the low-altitude radar-altimeter issues a command to break the metallic ribbons that attach the parachute-instrument module to the automatic Martian station. Simultaneously, the engine that controls the maneuvering of the parachute container is started.

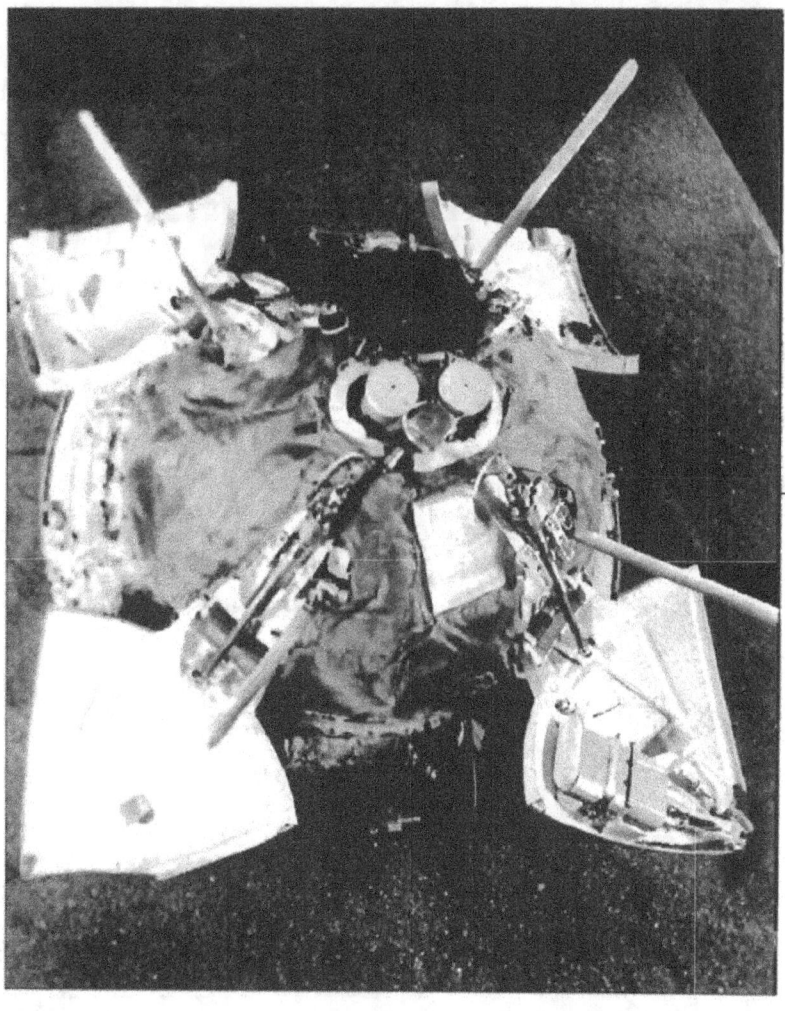

Figure II.

*Enlarged Diagram of the
Landing of the
Automatic Martian
Station of Mars 3--
(1) orbiter and lander
separation, (2) ignition
of the solid propellant
engine and transfer of
the lander from the flyby
to a Martian encounter
trajectory, (3) ignition of
the soft landing engine
and separation of the
engine with the para-
chute from the automat-
ic Martian station plus
withdrawal of the para-
chute, (4) landing of the
automatic Martian sta-
tion, inflation of the displacing bag, and separa-
tion of the aeroshell
cover, and (5) opening of
the petals, antennas, and
mechanisms, initiation
of the deployment of sci-
entific instruments on
the Martian surface, and
transmittal of informa-
tion to the Martian
satellite*

The automatic Martian station, which weights 358 kilograms, falls free to the Martian surface with a vertical speed of not more than 12 meters per second. In 15 seconds, after the station encounters the Martian surface, a program device gives a command to inflate the displacing bag. In 2 seconds, the expanded cumulative cartridge cuts the aeroshell cover. Under the impact of the compressed air, which was accumulated in the displacing bag, the aeroshell cover is thrown away from the station.

Thereafter, the PVM issues a sequence of commands to open the pyrotechnics lock, to ignite the cartridges in the four pyropushers, to begin the deployment of the scientific instruments that would study the Martian atmosphere and surface, and to start the onboard transmitters and telephotometers.

Data would be transmitted to the Martian satellite at a speed of 72,000 bits/sec in two independent radio channels. The circular panoramic images of the landing site would be transmitted as an image of 500 x 6,000 pixels. Each minute the video information would have to be interrupted by the telemetry data. The period calculated for each communication session would be 18–23 minutes.

After the first communication session, the onboard transmitters would be disconnected and the station transferred to a survival state. The second communication session would be initiated by a transmitted signal. Its period would depend on the position of the landing site and could be 0.7–5 minutes.

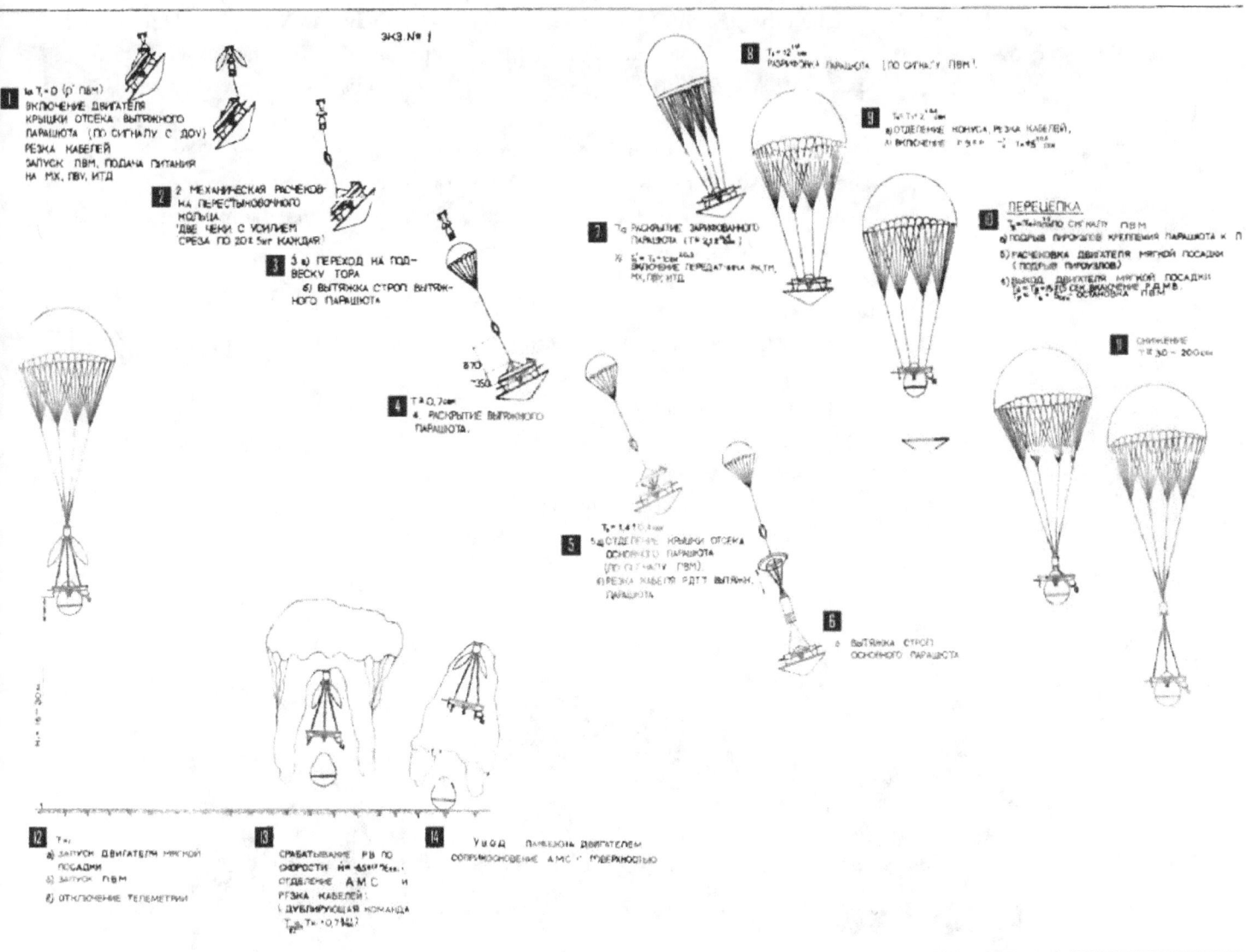

Figure 12.

The Landing Design (T1=0, the initiation of all landing operations)—(1) recorder of the relative accelerations issues a command at T1 to initiate a program timing device that controls the [...] the engine that activates the auxiliary parachute, (2) parachute with the cover is removed from the container, (3) shroud lines of the auxiliary parachute are pulled away, (4) auxiliary para[...] (5) at T2 (2.1 seconds), expanded cumulative cartridge cuts the parachute-instrument module, (6) top of the parachute-instrument module and attached cover with the main parachute are [...] lander, while shroud lines of the main parachute are pulled away, (7) main parachute is opened, and to decrease the overloading, its canopy is tightened with the rip cord, while at T3 (3.1 s [...] transmitters and scientific instruments are activated, (8) at T4 (12.1 seconds), rip cord is cut by the pyroknife, and canopy is opened completely, (9) at T5 (14 seconds), braking cone is sep[...] T6 (19 seconds), high-altitude radar-altimeter is activated, (10) at T7 (25 seconds), pyrolocks of the soft landing engine are ignited and engine is removed from the parachute-instrument m[...] T8 (27 seconds), low-altitude radar-altimeter is activated, (11) during descent with the parachute, landing engine is ready to be ignited, and predetermined descent time is 30–200 seconds, [...] distance of 16 – 30 meters from the Martian surface, and high-altitude radar-altimeter issues the commands to turn off the transmitters and scientific instruments and to ignite the soft lan[...] (13) speed of the lander descent is decreasing to 6.5 m/sec, and low-altitude radar-altimeter issues the command to separate the automatic Martian station from the parachute-instrument [...] (14) automate Martian station is on the Martian surface, and engine withdraws the parachute and parachute-instrument module from the automatic Martian station

4.5 Spacecraft Development and Testing

In February 1970, Babakin approved the preliminary design of Project M-71, which specified that the M-71C spacecraft should be launched in May 1971 (Figure 13). Its mission included the launch of the first Martian satellite and the launch of two spacecraft, Mars 2 and 3 (Figure 14), which were to be placed in a Martian orbit and deliver the first landers on the Martian surface. The M-71S spacecraft weighed 4,549 kilograms. This number included the weight of fuel and gas (2,385 kilograms). The Mars 3 spacecraft weighed 4,650 kilograms, including the weight of the lander (1,000 kilograms). To solve the specified scientific problems, the following instruments were to be installed on the spacecraft:

(a) Fluxgate magnetometer
(b) Infrared radiometer to study the distribution of the temperature on the Martian surface
(c) Infrared photometer
(d) Spectrometer to determine the concentration of water vapor in the Martian atmosphere,
(e) Photometer working in the visible part of the electromagnetic spectrum to study the reflectivity of the Martian surface and atmosphere

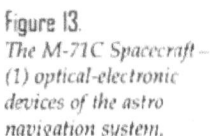

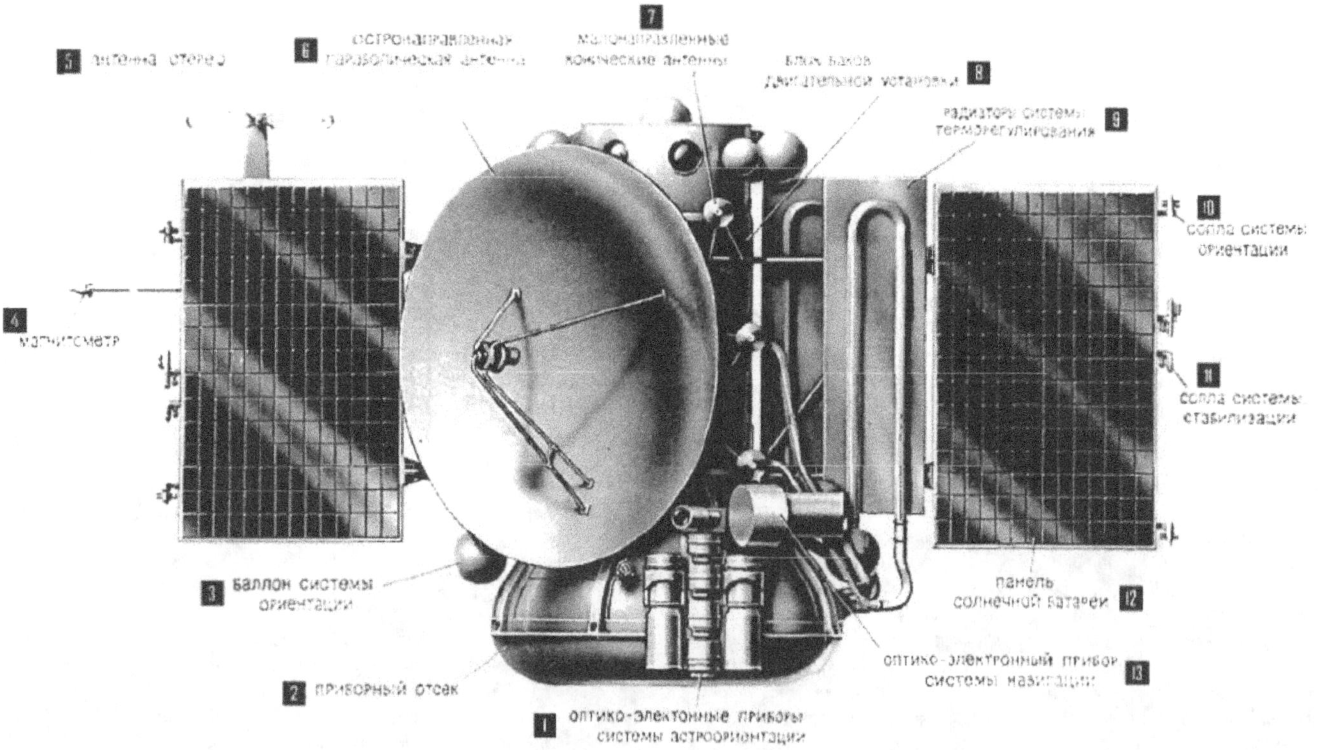

Figure 13.
*The M-71C Spacecraft –
(1) optical-electronic devices of the astro navigation system,
(2) instrument module,
(3) container of the attitude system,
(4) magetometer,
(5) stereo antenna,
(6) high-gain parabolic antenna, (7) low-gain cone antennas,
(8) tanks of the propulsion system, (9) radiators of the temperature control system,
(10) nozzles of the attitude system,
(11) nozzles of the stabilization system,
(12) solar panel, and
(13) optical-electronic device of the positioning system*

Рис.   КОСМИЧЕСКИЙ АППАРАТ «М-71С»

(f) Radiometer to study the dielectric permeability and the temperature on the Martian surface

(g) Ultraviolet photometer to study the Martian upper atmosphere

(h) Cosmic ray particles detector

(i) The detector to determine the kinetic energy of electrons and protons, the charged particles spectrometer

(j) Telephoto cameras with focal distances of 52 and 350 millimeters

In addition, the M-71S spacecraft and Mars 3 were equipped with stereo systems to support the joint Soviet-French experimental program to study solar radiation at the frequency of 169 MHz. The total weight of the scientific instruments installed in the Mars 3 spacecraft was 89.2 kilograms.

The landers of the spacecraft Mars 2 and Mars 3 carried the instruments to measure the Martian atmospheric pressure, the temperature, and the wind velocity, to define the chemical composition of the Martian soil, and to study the physical-mechanical properties of the soil's surface layer. In addition, two telecameras for making panoramic images of the landing sites were installed. Each lander contained 16 kilograms of scientific instruments.

The time remaining before the launch of the spacecraft was limited. To succeed, each stage of the project had to be completed in the shortest time. On the other hand, this approach should not degrade the quality and reliability of the spacecraft.

At that time, information about a new concept of space project management, developed in the United States, became available. The mass media indicated that the new management concept allowed for a reduction in the time for developing and building the ballistic rocket Polaris.

In my view, the new concept was very attractive. At each stage of the

Figure 14.
*The Mars 3 Spacecraft—
(1) instrument module,
(2) high-gain parabolic
antenna, (3) stereo antenna of the scientific instruments, (4) lander,
(5) low-gain antennas,
(6) radiators of the tem-
perature control system,
(7) solar panel, (8) reactive nozzles, (9) device for
planetary angle measure-
ments, (10) Earth sensor,
(11) star sensor,
(12) precise solar sensor,
and (13) rough solar
sensor*

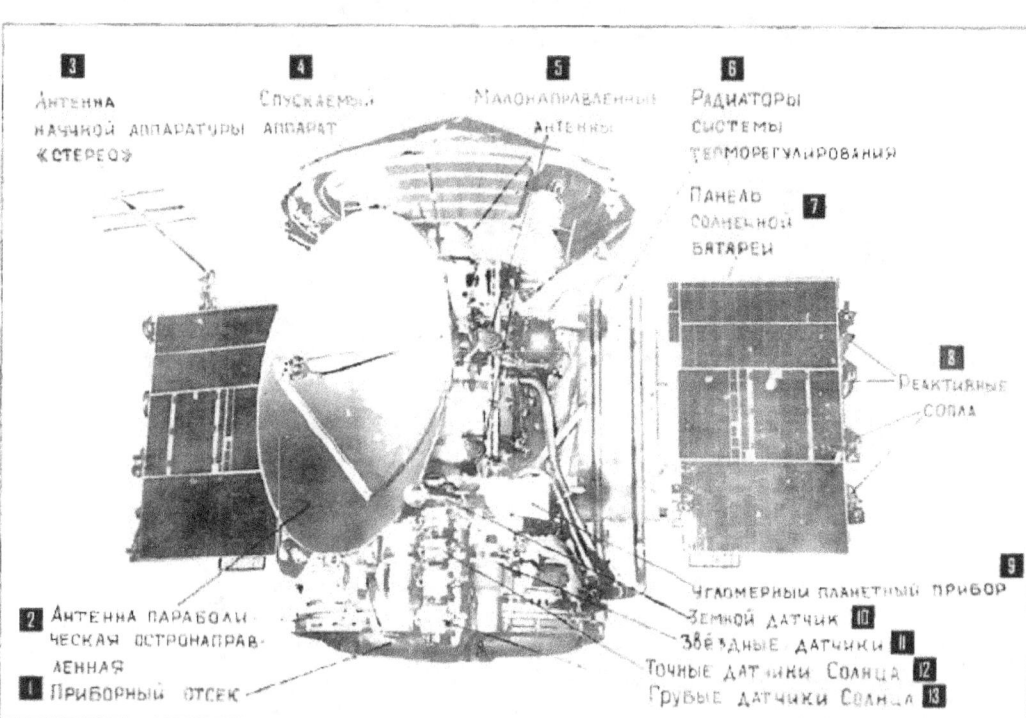

project, the time limitations for each job were clearly outlined. At this time, Babakin appointed me his deputy for the development of spacecraft for the exploration of Mars and Venus. In addition to the technical problems, I was in charge of project assignments that should be finished on time and with a high quality. To reduce the time needed to develop and publish the design documentation, I decided to use the American system of planning and management. More than 15 days of exhausting work were required to outline the preliminary version of the management structure. The same time was needed for the administrative department to make some corrections, remove circular connections, to code events, and so on. As a result, a poster 3.5 meters in length, filled with events, connections and other information, was produced.

In November 1970, the management plan was approved by Babakin, and in June 1971, the technical documentation was issued. The skeptics who believed that the American management concept could not be used in Soviet Union were embarrassed. Minister Afanasiev instructed the Urals-Siberian plants to deliver the parts for the spacecraft and landers to our organization. As a result, we could soon initiate the complex program of testing operations.

The testing of the lander and its system consumed a lot of time. Fifteen launches of the meteorological rocket M-100B were performed to test the initiating and opening of the parachute system under different flight conditions. During testing operations, it was discovered that the main parachute had a tendency to collapse.

Figure 15.
The General Scheme of the Catapult- (1) catapult, (2) platform with the duplicate soil, (3) protective screen, (4) lock, (5) winch, (6) rope, (7) sledge-thermostat, (8) automatic Martian station, (9) top of the thermostat, (10), rope, (11) elastic shock absorber, (12) stopper, and (13) trajectory of the automatic Martian station after it hits the platform (α—angle of catapult slopes inclination, β—angle of the platform inclination)

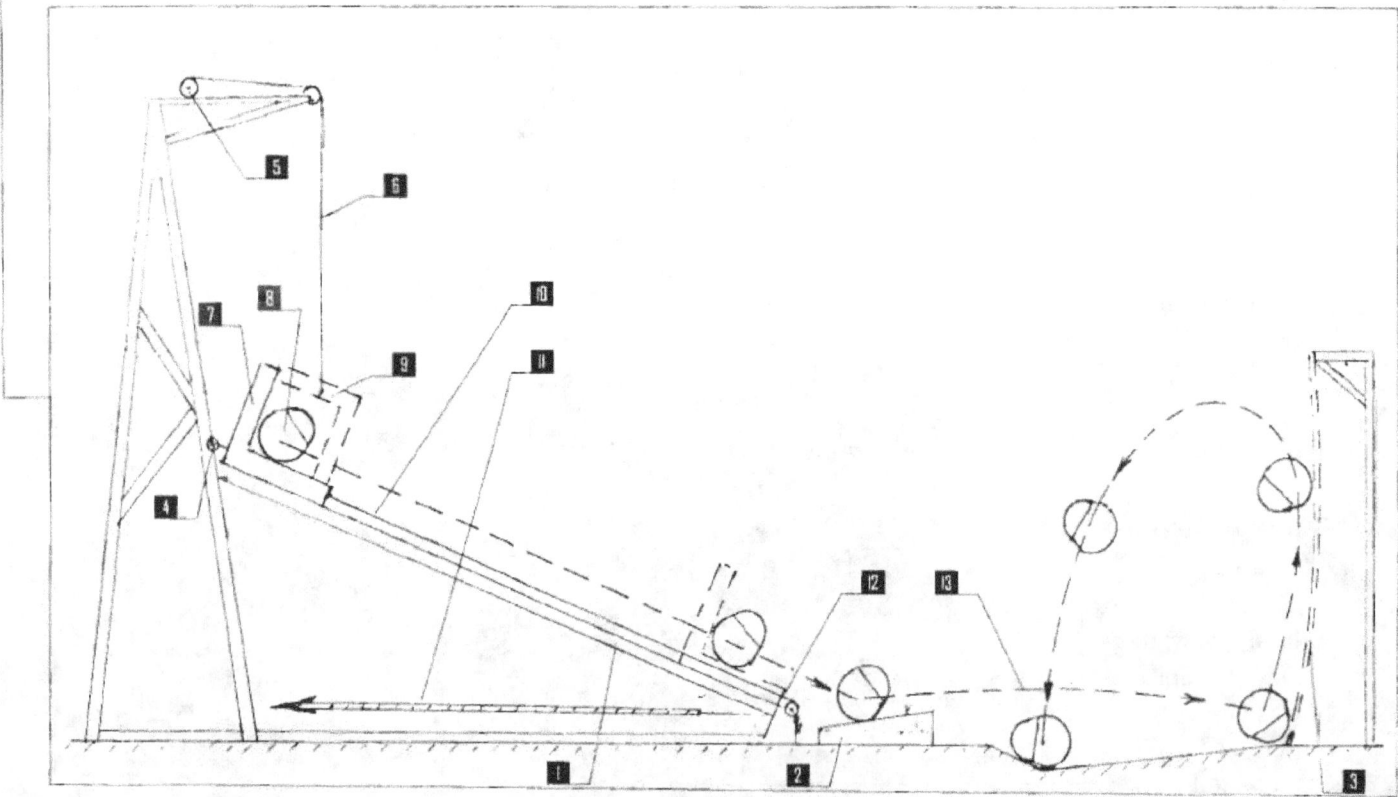

To eliminate this trend, it was necessary to reconsider the design of the parachute system. Different pyrotechnic systems, such as expanded cumulative cartridges, pyrolocks, pyropushers, pyrocylinders, and so on, were broadly used in the design of the spacecraft and lander. Pyrotechnic systems were tested extensively because, according to the standards established in the Lavochkin design bureau, the pyrosystem could be installed in the spacecraft only if it worked 12 times in different conditions without failures in an Earth-based facility.

To test the soft landing systems, five lander test models equipped with radio-altimeters, onboard automatic units, and gunpowder engines were dropped by parachute. The parachute system, with a reduced area of canopy, was used to provide the predetermined vertical descent speed for the lander test models. During all drop experiments, the soft landing system operated properly.

A special testing facility was used to check the protection efficiency and operation of the automatic Martian station after its impact with the rocky soil (Figure 15). The testing facility was made up of (1) a catapult, (2) a duplicate of the Martian soil, and (3) a wall designed for protection. Lander test models equipped with simulations of the onboard instruments, the standard pyrosystems, a program timer device, and a power system were dropped by parachute five times. While the parachute drops were performed, the temperature of the foam plastic cover varied from +50 degrees Celsius to –50 degrees Celsius. Simultaneously, the pitch angle varied from 0 to 180 degrees, and the angle of incidence to the duplicate soil changed from 0 to 10 degrees.

During testing, horizontal speeds of 28.5 meters per second and vertical speeds of 12 meters per second at the time of landing were simulated. The tests showed that foam plastic provides a proper shock absorption and protection of the capsule during landing and if the G-force does not exceed 180 units. The separation of aeroshell cover and the placement of the station in a vertical position under maximum speeds and with different directions of impact were accomplished properly.

In conclusion, the simulation of all flight conditions was made. The operational ability of the lander was checked in different stages of the simulated flight. Vibration and linear overloading tests simulated the launch of the spacecraft and its placement in an interplanetary trajectory. Centrifuge tests simulated the ballistic braking of the lander, and the drop of the automatic Martian station was simulated by the catapult with the load factor of 180 units. After the above simulations, the onboard program-timing device was initiated. By its command, the aeroshell cover was separated, the flaps were opened, and the station was placed in a vertical position.

Simultaneously, the transmitters and scientific instruments were turned on, and the x-ray spectrometer was placed on a duplicate of the Martian soil. The PROP-M instrument made its short trip, still analyzing Earth's soil. During the next 25 minutes, panoramic images and scientific data were transmitted and received by the radio system. In addition, for 25 hours, the station was tested in a vacuum chamber under a pressure of 6 mbar while being exposed to airflow with the speed of 25 m/sec. Daily variations of the temperatures of the Martian atmosphere were simulated as well.

Then the testing operations were completed, and a second communication session was started. A whole cycle of testing operations was made without any failures, and we became confident that the automatic station could accomplish the tasks planned for it.

The utilization of the lander in Project M-71 required that it be sterilized. The development of a sterilization procedure was assigned to Academician A.A. Imshenetskiy. The Institute of Sterilization and Disinfection of the Health Protection Ministry was responsible for sterilization procedures.

A few ways were proposed for the sterilization, namely, gaseous, utilizing methyl bromide, radioactive and thermal. None of these techniques could be used with the completely assembled lander. However the lander's parts could be independently sterilized by any of these techniques. To preserve the sterility of the lander during its assembly, a special sterile facility with a sluice chamber was built. This facility was equipped with filters through which air was transferred under low pressure. This facility was equipped with the bactericidal lamps.

4.6 The Spacecraft's Struggle to Mars

May 1971 had arrived. Recalling our trouble with the M-69 spacecraft, we nervously awaited for three upcoming launches. Spacecraft M-71 was launched on May 5, 1971. At first, the ignition of all three stages of the Proton rocket and booster block D worked properly. The spacecraft with booster block D was placed in the predetermined Earth orbit.

In approximately 1 hour, after orbiting Earth and approaching Guinea Bay, block D should have started a second time and transferred the spacecraft to

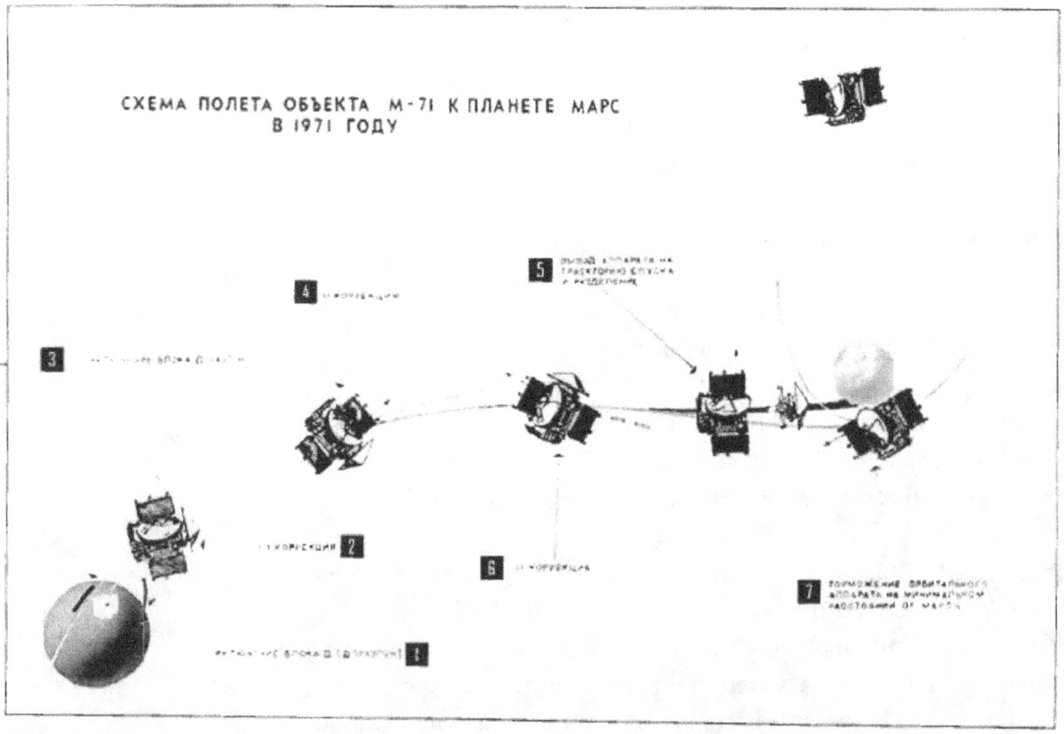

Figure 16.
The Flight Profile of the Mars 2 and 3 Spacecraft to Mars in 1971— (1) activation of block D (before boosting), (2) first correction, (3) activation of block D (boosting), (4) second correction, (5) injection of the spacecraft in the descent trajectory and separation, (6) third correction, and (7) braking of the orbiter at the minimal distance from Mars

СХЕМА ПОЛЕТА ОБЪЕКТА М-71 К ПЛАНЕТЕ МАРС В 1971 ГОДУ

an interplanetary trajectory. A ship in Guinea Bay did not show that the engines of block D had ignited. In a few minutes, block D flew over the Yevpatorijan (Crimea) control facility. The telemetered data showed that no command was issued to start the engine. An analysis showed that the operator had made a mistake. He issued an eight-digit code command to the spacecraft control system to activate block D for the second time in a reverse order. It was a human factor, not technology, that was responsible for the error.

We lost the opportunity to launch the first Martian satellite. Besides that, we lost the radar beacon that provided information about the position of Mars in space. The flight profile of the preliminary design (Figure 16) could not be accomplished because precise data on the Mars position in space had not been acquired. These data were needed to calculate the angles at which the lander should enter the Martian atmosphere. Now the only hope left was that the space automatic positioning system would not fail. This alternative system was designed in case the M-71S spacecraft would fail.

A decision about its development was made in the summer of 1970 at the Council of Main Designers. The system used an optical angle measurement instrument, which was developed under the supervision of the main designer V.I. Kurushin. About 7 hours before reaching Mars, the instrument should perform the first measurement of the angular position of Mars with respect to the base coordinate system (Figure 17).

Figure 17.
The Flight Profile of the Mars 2 and 3 Spacecraft Near Mars—(1) first measurement, (2) third correction, (3) lander separation, (4) lander braking, (5) pitch and yaw, (6) second measurement, and (7) orbiter braking

(8) Height of the pericenter:
Before correction—
2,350 ± 1,000 km
After correction—
1,500 ± 200 km
Distance to Mars:
First measurement—
~70,000 km

Second measurement—
~20,000 km

Speed Impulses:
Correction—≤100 m/sec
Lander braking—100 m/sec
Orbiter braking—
1,190 m/sec
Time of lander descent—
~6 hours

Scatter of entry angles for the lander: ±5 degrees

Scatter of rotation period of the orbiter: ±2 hours

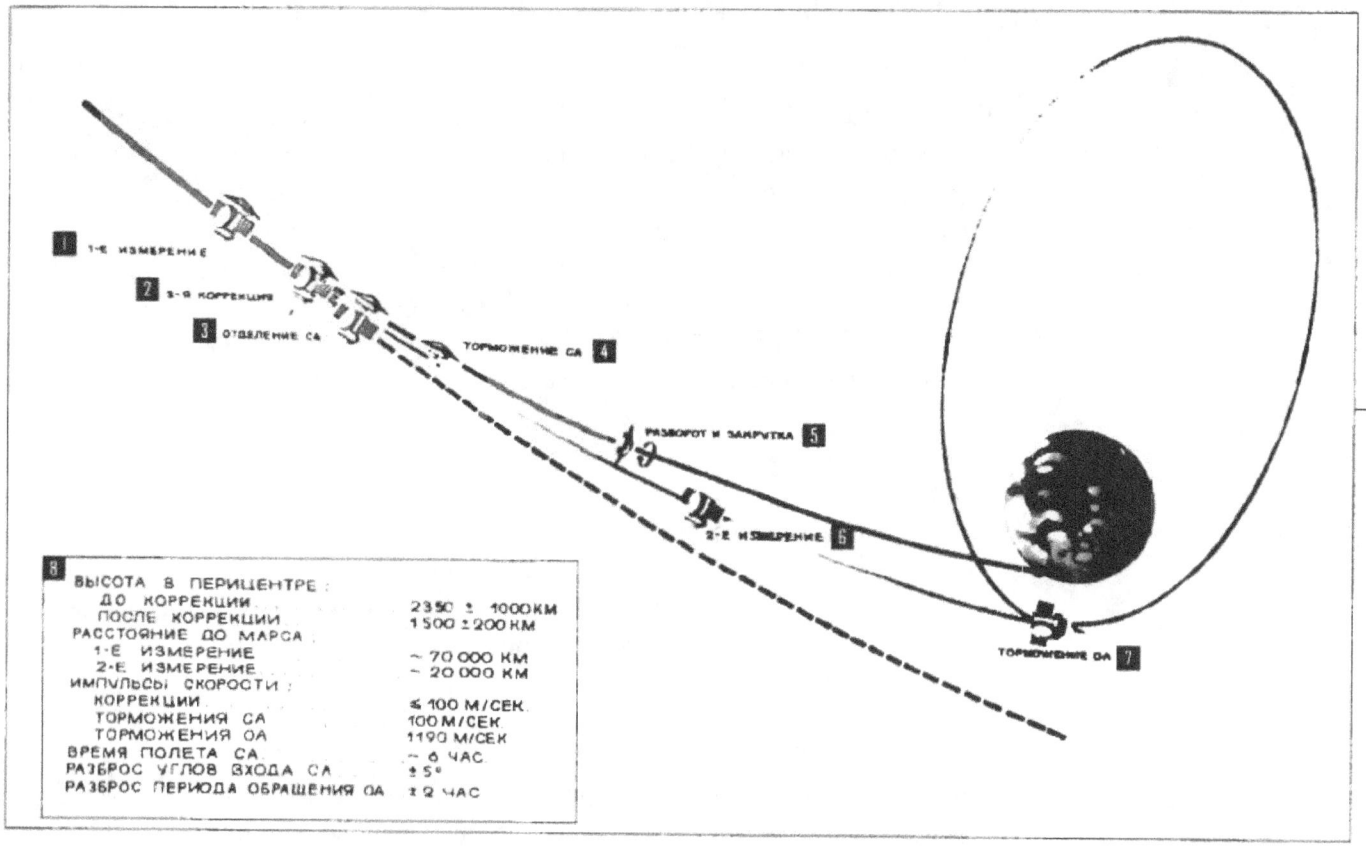

The measured data should be transferred to the onboard computer of the control system, which calculated the vector for a third correction needed to relocate the spacecraft to a nominal trajectory. In accord with the results of calculations, the control system should issue the following commands:

1. Turn the spacecraft to a position required to perform the correction
2. Start the engine for the correction braking propulsion system, and stop the engine after the correction impulse is terminated
3. Turn the spacecraft to a position required for the lander to be separated and to separate a lander
4. Rotate the spacecraft about three axes using the Sun and star Canopus as reference
5. When the spacecraft approaches the planet, at the distance about 20,000 km, conduct the second measurement of the Mars angular position with respect to the base coordinate system

As a result of the second measurement, the onboard computer had to calculate the vector direction of braking impulse to relocate the spacecraft in the Martian satellite orbit. These operations should be performed automatically without commands from Earth. The test of the optical angle measurement instrument was performed at the control system testing facility and was completed successfully. Only 2 to 3 months were allocated to testing the computer programs. This limited amount of time caused concern.

On May 19 and 21, 1971, the spacecraft Mars 2 and Mars 3 were launched in an interplanetary trajectory. All stages of the Proton rocket and booster blocks D operated perfectly. In accord with the results of the outbound trajectory measurements, the first correction of the Mars 2 and Mars 3 spacecraft trajectory was made on June 5 and on June 8, 1971, respectively. The onboard equipment of both spacecraft performed without any problems. The group of flight managers started their routine work. Because of the particular flight trajectories to Mars, communication with the spacecraft was maintained at night.

After the communication sessions were completed, the management group analyzed the data acquired, then if it was necessary reconsidered the agenda for future communication sessions, and transferred the report of their work to Babakin. Then the group rested. Every morning, upon coming to work, we listened to the presentation of V.G. Timonin, who was the head manager. If it was necessary, we instructed the OKB departments to undertake the jobs required.

A bright sunny morning on June 25 did not forecast any troubles. The telephone on Babakin's dedicated line rang unexpectedly. "Come urgently," his troubled voice said. Gloomy Babakin was pacing his office. "I just spoke with Timonin," he said. "Communication with both Mars failed last night. Get the team together and fly out to Yevpatorija immediately."

I was amazed by Timonin's news. The defects of spacecraft Mars 2 and Mars 3 were exactly identical. The original transmitters, which work in the decimeter band, failed first. Then the duplicate transmitters were activated. In the beginning, they worked properly but later on failed as well. A command was given to start the transmitters working in the centimeter band. Telemetry data indicated that the centimeter band transmitters were activated and started to work, but the signals did not reach Earth. To avoid the development of an emergency situation, the transmitters were turned off, and we awaited directions from Moscow.

I telephoned Yu.F. Makarov, who was the assistant of Ryasanskiy and was in charge of the onboard radio system. He was already aware of the situation. We agreed that they will have a team of four people and we will have a team of three people.

In the middle of the vacation season, it was impossible to buy a ticket to Simferopol at the Aeroflot terminal. I attempted to buy tickets reserved for the staff of the Council of Ministers. Unfortunately, only tickets with a departure date of 1 week were available. From the Kremlin, I telephoned G.A. Tyulin, who was the deputy minister. I explained to him the situation with the Mars spacecraft and with the Aeroflot tickets.

Tyulin gave us permission to use the ministry's airplane. The airplanes of our ministry were parked at Vnukovo-3 airport. Like all passenger aircraft, they had Aeroflot written on their side.

In the morning of June 26, 1971, we flew out to Simferopol on airplane AN-24. During the flight, we discussed the information received and outlined the future plan of work. Finally, we arrived at Simferopol. The airplane landed and moved to the terminal. We proceeded down the stairwell and went to the exit.

Porters with their carts rushed toward us in a chaotic fashion. Upon exiting the airport, we turned around and observed a silent scene like in Gogol's play *Inspector*. Stunned porters are struggling to grasp the reality. In the middle of the vacation season, an empty airplane had arrived from Moscow.

Soon we reached the flight control facility and carefully analyzed the telemetry information. Strangely enough, there was no sign of the abnormal functioning of the telemetric system during the previous communication sessions. We discussed a suggested plan of actions with the management team and introduced some changes based on their comments.

At night, when the Mars spacecraft became visible, we began its implementation. The first night went by. A few straws of hope appeared for reestablishing communication in the decimeter band. A whole day was devoted to analyzing the telemetric data acquired at night and developing a new plan. The second night was productive. We identified the operating conditions in which the duplicate transmitters worked in the decimeter band without errors. Simultaneously, communication with

S.S. Kryukov, main designer of the Lavochkin design bureau

the original transmitters was reestablished, although the time of their operation was limited. However, the signal from the transmitters operating in the centimeter band did not reach Earth. The radiation sensor that was installed at the antenna's mirror permanently showed nothing. We could not find an explanation for this phenomenon.

The idea that the antennas absorbed solar energy, which heated and melted the radiator, was not justified by calculations. The antenna was directed to the Sun for no more than 10 seconds. The radiators were manufactured using silver solder whose melting temperature was 700 degrees Celsius. Nevertheless, in further projects, we installed a cloth tent above the antenna's mirror. The antennas' design did not change, but they operated perfectly with the tent.

The exhausting work was completed. For 2 days, we had not slept at all and now could get some rest. At that time, Minister Afanasiev was at the flight control facility. Probably, his visit was connected with the flight of the Soyuz 11 spacecraft.

The next day, Afanasiev asked me to present the results of our effort. Our actions were approved. Immediately after my presentation, the communication session with the Soyuz 11 spacecraft was initiated. By the loud speakers, the troubled voice of cosmonaut V. Volkov said, "It is impossible to eat the steaks," and in few minutes, his voice was heard again, "Fire in the spacecraft." I believed that his excitement could be explained by the fact that he predicted the danger. The head manager of flight asked, "Commander, what's happening on board?" G. Dobrovolskiy reported, "We had a short circuit but have already fixed it; Volkov is a little excited. Don't worry, everything will be all right." The next day, the team of cosmonauts of the Soyuz 11 spacecraft crashed during their flight back to Earth. Perhaps, some human beings can predict the danger. I recollect a similar example.

In March 1968, in the headquarters of the Council of Ministers, I met Yu.A. Gagarin. He was slightly overweight and wore a military greatcoat, which obviously did not fit him. Probably, he was waiting for somebody. At this time, the VPK Resolution, which determined the launch of the spacecraft Venus 5 and Venus 6, was almost completed. I told him jokingly, "Yuriy Alekseevich, we have the position of the main pilot vacant on the Venusian spacecraft." He did not respond to my joke and with gloomy smile said, "Maybe next time." In a few days, Yu.A. Gagarin died.

On July 1, we returned to Moscow and as usual began to work. We did not anticipate the future loss. Very soon, we were stunned with the news that Babakin died from a heart attack. The death of Babakin was an enormous loss for space technology and for our team as well. Nobody could replace him. Talented people are not born very often.

S.S. Kryukov, the former first deputy of Babakin, was appointed the main designer. Kryukov was a highly qualified specialist. He used to be the deputy of Korolev and was in charge of the development of the rocket that launched the first Earth satellite. In 1970, Kryukov got a position in the Lavochkin design bureau. The Babakin-Kryukov team was coordinated well and worked very productively.

The flight of spacecraft Mars 2 and 3 continued. In November 1971, a second correction of their trajectory was made successfully. A few days remained before the spacecraft would approach Mars. At that time, the Martian weather was not good for observations with the orbiter and especially with the lander. For the last few weeks, an unusually strong dust storm covered the whole surface of the planet. Astronomers indicated that such a large dust storm had never before been recorded on the Martian surface.

On November 21, 1971, the space automatic positioning system was used to make the third correction to the trajectory of the Mars 2 spacecraft trajectory. The Mars 2 spacecraft lander was directed to the planet. At the same time, the spacecraft was placed in a Martian orbit with a pericenter height of 1,350 kilometers and a period of rotation of 18 hours. This operation was not successful. Apparently, after the second correction, the Mars 2 spacecraft trajectory had been close to the predetermined trajectory.

Nevertheless, the onboard computer issued a wrong command to decrease the height of the pericenter of the flyby hyperbola. As a result, the lander entered the Martian atmosphere at the big angle and hit the Martian surface before the parachute system was activated.

Because of time limitations, the testing of the computer programs for the space automatic positioning system was not completed. We just did not have time to perform computer modeling of a situation in which the real and previously calculated trajectories would coincide.

Could we have avoided these results? Definitely yes, but only if the first space barter in the history of human civilization would have happened 1 year earlier, in 1971. The barter included the exchange of the American precise data for the Martian ephemeris for the Soviet data on Venus. If that had taken place and being aware that the spacecraft was close to its predetermined trajectory for a Mars flyby, we would have activated the Earth-based control system and the flight would have been controlled by commands from Earth.

A third correction to the Mars 3 spacecraft trajectory was made on December 2, 1971. The lander separated from the spacecraft and entered its predetermined trajectory for a Mars encounter. After 4 hours and 35 minutes, the lander entered the Martian atmosphere and landed at a location with coordinates 45° S and 158° W.

The spacecraft was placed in a Martian satellite orbit with a pericenter height of 1,500 kilometers and a period of rotation of 12 days and 19 hours. The predetermined period of rotation was 25 hours. The discrepancy between the real and the predetermined periods of rotation could again be explained by the time limitations, which did not allow for proper testing of the computer programs developed for the space automatic positioning system.

If the speed of the spacecraft changed rapidly, the control system miscalculated the impulses transferred by the gyro-integrator and issued a command to stop the engine of the correction braking propulsion system. That happened irregularly, usually before or after the predetermined impulse of the fast moving spacecraft had been exceeded.

Everybody impatiently waited for the information transmitted from the lander to the radio system onboard the spacecraft. Using the high-gain antenna, the information recorded on magnetic tape was transmitted to Earth when the spacecraft was in the state of three-axis stabilization.

To relocate the spacecraft to a position of three-axis stabilization, it was necessary to open the cover of the star sensor, which protected it from contamination by products of burning. According to calculations, the cloud of the burned products should disappear 30 minutes after the engine was turned off. The cover of the star sensor was opened. In a few minutes, the three-axis stabilization was lost. Dirt was deposited on the viewport of the Sun sensor.

After an hour, the cover of the second star sensor was opened, and the spacecraft was transferred to a position of three-axis stabilization. The transmission of panoramic images of the Martian surface recorded on the magnetic tape was initiated. The main engineer of NII KP, Yu.K. Khodarev, who was standing close to the rack where the signal was displayed, gave a command to reduce the signal because it was too strong. Then, the telephotometer data were transmitted. There was a gray background with no details.

In 14.5 seconds, the signal disappeared. The same thing happened with the second telephotometer. Why did two telephotometers working in independent bands simultaneously fail within a hundredth of a second? We could not find an answer to this question.

Later on, I discovered an interesting fact. During the second world war, British radio operators had their transmitters malfunction because of a coronal discharge while working in the desert of Lebanon during a dust storm. On Mars, the size of dust particles, the humidity, and the atmospheric pressure are much less, but the wind velocity is much higher than in the desert of Lebanon.

Perhaps, the coronal discharge was the reason why the signal from Mars suddenly disappeared. Even if the radio link worked properly, in a dust storm, we still would be unable to get the topography images. It was planned that after a 1-minute time interval, the transmission of the panoramic images would be followed by the transmission of data on the physical-chemical parameters of the Martian soil. It was especially valuable for science that data on the atmospheric pressure, temperature, and wind velocity at the Martian surface be acquired in conditions of an unusually strong dust storm.

The dust storm on Mars continued to rage. Both Mars orbiters continued to make images of the Martian surface, but the dust completely obscured the topography. Even the mountain Olympus, which had an elevation of 26 kilometers above the average Martian surface, was invisible. During one of the mapping sessions, the image of the full Martian disk with a distinct layer of clouds above the dust cover was obtained.

Transmitters that worked in the centimeter band failed. That was the reason why the scientific data and images from the Martian surface were transmitted in the decimeter band by the transponders that were "cured" in June. As usual, the main designer of the radio system, Ryasanskiy, had additional resources, but at this point, he did not speak of them.

For the radio systems of the Mars 2 and Mars 3 spacecraft, he generated a PN code, which allowed data to be transmitted in a special mode. That particular mode made it possible to increase the speed of transmission in the decimeter band. This allowed the transmission to Earth of all scientific data acquired by the Mars 2 and 3 spacecraft for the 8 months of their existence and completely accomplished the program of Mars exploration with the orbiters.

4.7 Main Results of Mars Exploration With the Mars 2 and 3 Spacecraft

The Mars 2 and 3 spacecraft were equipped with scientific instruments designed for the remote and planet-based study of the Martian surface and atmosphere.

As a result of this study, the temperature and atmospheric pressure on the Martian surface and the nature of the surface rocks along the tracks of the satellite orbits were defined. The data on the soil density, its heat conductivity, dielectric permeability, and reflectivity were acquired. The temperature in the lower atmosphere and its change with time and latitude were determined. The heat flow anomalies on the Martian surface were discovered.

Also, it was defined that the temperature of the northern polar cap was less than –110 degrees Celsius. Altimetry data along the satellite's tracks were obtained. It was defined that in the Martian atmosphere, the concentration of water vapor was 5,000 times less than in Earth's atmosphere. The data on extent, composition, and temperature of the upper atmosphere were obtained. The altitude of the dust clouds and the size of dust particles were determined.

The Martian magnetic field was measured. Colorful images of Mars were obtained. The optical compression of the planet has been refined. The layered structure of the Martian atmosphere, its luminosity at a distance of 200 kilometers beyond the Mars terminator, and the change in its color close to the terminator were discovered.

One of the most important accomplishments of Project M-71 was that the scientifically and technically intricate problem of a soft landing on the Martian surface was solved.

"The Russian Armada is moving towards the red planet." That was the headlines of many newspapers in August 1973. There certainly were reasons for newspapers to smell a sensation. In the Soviet Union and in the United States, two identical spacecraft, the main one and a duplicate, are usually launched. Because of the space race, this strategy was justified because the program should be carried on even if one of the spacecraft broke down. The world community was stunned by the fact that the Soviet Union launched four spacecraft. Scientists believed that the Soviet Union decided to make a powerful thrust in the exploration of Mars and solve many Martian problems. But the underlying reason for the launch of four spacecraft was more political in nature.

In 1975, the United States planned to deliver the Viking lander to the Martian surface. Scientifically, the Viking lander was much more advanced than the M-71 lander. Being aware of the potential failures of the M-71 spacecraft, the government made a decision to launch the spacecraft in 1973, before the results of the Project M-71 would be available.

However, in 1973, the ballistic conditions for flights to Mars were not good. The speed required to place a spacecraft in an interplanetary trajectory had to be increased. In addition, it turned out to be impossible, as required in Project M-71, to deliver the spacecraft with the lander to an orbit that would allow the descent to be slowed by one braking rocket.

The only solution that would be in accordance with the government resolution was to launch four spacecraft. Two of them would become Martian satellites and two other would deliver landers to the Martian surface. Their general design and the onboard systems were identical to the Mars 2 and 3 spacecraft.

However, a number of the scientific instruments had to be redesigned. Simultaneously, an additional radio system had to be installed on the lander so it could transmit to Earth the information acquired after the parachute started to be deployed. Apparently, this system increased the probability of receiving the data transmitted from the lander.

Yu.F Makarov, the deputy of the main designer of NPO KP, supported my idea to install a new radio system in the lander. When the system was completed and was placed in the lander, there was a "kind soul" who convinced the main designer that my solution was flawed. An unpleasant conversation took place with Kryukov. He accused me of the installation of a system that, in his opinion, had no value.

However, I was consoled later, when the telemetry data from the Mars 6 lander were transmitted exactly in this radio band. The technological documentation used in the new project, except the theoretical calculations connected with the specific date of the launch, was the same as for Project M-71. Beyond that, the joint effort to deliver parts and onboard systems was well coordinated. All of that gave hope that the spacecraft would be built on schedule.

There was concern about the concentration of four spacecraft simultaneously at the testing control facility (KIS). Four work places had to be organized. However, even with the extended 12-hour shift, at least eight teams should work. We realized that this would be a difficult stage, but it was not as difficult as we had thought.

Again, Mars challenged us. Suddenly, during testing of the power system, the onboard blocks started to fail. Analysis showed that in all cases the power system malfunctioned because of a failure of the 2T-312 transistor, which was fabricated at the Voronezhskiy plant. All blocks of the onboard equipment were literally filled with these transistors. An interministry commission carefully analyzed this problem and came to the conclusion that the reason of the transistor's failure was intercrystalline corrosion in the area of the transistor lead.

To save the gold resources, some "smart person" suggested that the gold leads be replaced by aluminum ones. The necessary tests were not made. And so 2 years later, this suggestion caused major trouble. The only way to remedy this situation was to replace the flawed transistors with transistors fabricated according to the old technology.

An experimental study showed that this task would require at least 6 months. This was exactly the time remaining until the spacecraft should be launched. To find a solution for this dead-end problem, A.N. Davydov, the head of the reliability department of the Lavochkin design bureau, was ordered to conduct a careful analysis of all known failures of the 2T-312 transistors, including failures in instruments of other organizations. Also, he was directed to consider the results of this analysis as it might affect the reliability of spacecraft already using this equipment.

The analysis showed that the transistor's failure rate begins to increase 1.5–2 years after its production date. For transistors installed in our instruments, that corresponded to the arrival date of the spacecraft to Mars. The probability of completing the flight program, while taking into account the possible failure of the 2T-312 transistors, was estimated at 0.5. Based on this information, the ministry officials made the decision to continue with the launch of the spacecraft as previously planned.

Testing of the spacecraft's power system proceeded at a full pace at KIS. Suddenly, another late night telephone call. The nervous shift supervisor informed us that they had an emergency and asked us to come over there urgently. There, I saw a very unusual scene. The control system was turned off, and gloomy operators were aimlessly wandering around the facility. We found that during the general testing operations, when the cover of the instrument module was opened and the onboard instruments could be easily accessed, a particular testing of the control system was required.

Technicians joined the connectors, and according to the technical procedure, the power system was turned on. Then the control system was tested for 20 minutes. However, soon it was discovered that the information obtained was not in accordance with the program commands. The program was stopped, and the circuit design was checked. It was found that the technician managed to join a hundred terminal connectors in a wrong way.

In spite of this obvious mistake, the connection had been approved by the department of technical control and by the military representative. That was clear by the records and signatures in the technical documentation. What happened; why had the triple control not worked? That was the rhetorical question.

I telephoned G.M. Priss, the deputy of Pilyugin. At NPO AP, he was in charge of the control system of the Martian spacecraft. I told him all the details of our emergency situation. For a while, Priss reflected and after that suggested that we drive over to NPO AP. We grabbed the test results and raced through the deserted Moscow night. During the trip, we developed a plan. We suggested that the control system be removed from the spacecraft and installed and tested in the testing facility of NPO AP. Only after the main designer and the military representative had approved the results of the test would the control system be installed in the spacecraft again. To reduce an inevitable delay in the testing of the power system, we suggested that the order in which spacecraft would be launched be changed. Specifically, we suggested launching the spacecraft where the emergency situation took place the last time.

Eventually, we arrived at NPO AP. We passed the empty and resonant corridors and went upstairs into the office of Priss. We were awaited, and the flow charts of the control system were laid on the desk. Priss had already discussed the possible solution of the problem with specialists from NPO AP. During the last session, parameters of the control system were recorded. We put the notes with records on the desk. The specialists bent over the records and sometimes made short comments that I could hardly understand. That reminded one of a consultation near the patient's bed when doctors discuss the diagnosis and how to cure the disease.

Eventually, the meeting finished. I presented our suggestions to recheck the control system in the testing facility of NPO AP. However, the decision of Priss completely changed our plans. He indicated that analysis did not reveal any damage to the blocks of the control system. In addition, he suggested urgently, before the day shift would start to work and the military representatives of NPO AP show up, to place the control system in the spacecraft and test it again. He was concerned that being aware of the emergency situation, the conservative military department of NPO AP could veto the use of this control system and the launch of the spacecraft would not be feasible.

After our departure, the specialists from the Lavochkin design bureau dispersed and went home. Only one person, the director of this organization, A.P. Milovanov, could get them together. I telephoned him, excused myself for the night call, and briefly explained what happened and what decision had been made. He immediately understood the situation, asked me to calm down, and promised that all specialists would be at their workplaces when we come back to the Lavochkin design bureau.

We raced through the Moscow night again. Priss did not accompany us. He did not want to show up and provoke unnecessary questions. He was a very smart man. Only a person who was aware of all details of the design of the control system, who trusted the specialists and who was deeply concerned with the fate of the project,

Figure 18.
The Mars 4 and 5 Spacecraft – (1) high-gain antenna, (2) radiometer, (3) fuel tank, (4) radiators of the temperature control system, (5) solar panel, (6) instrument module, and (7) magnetometer

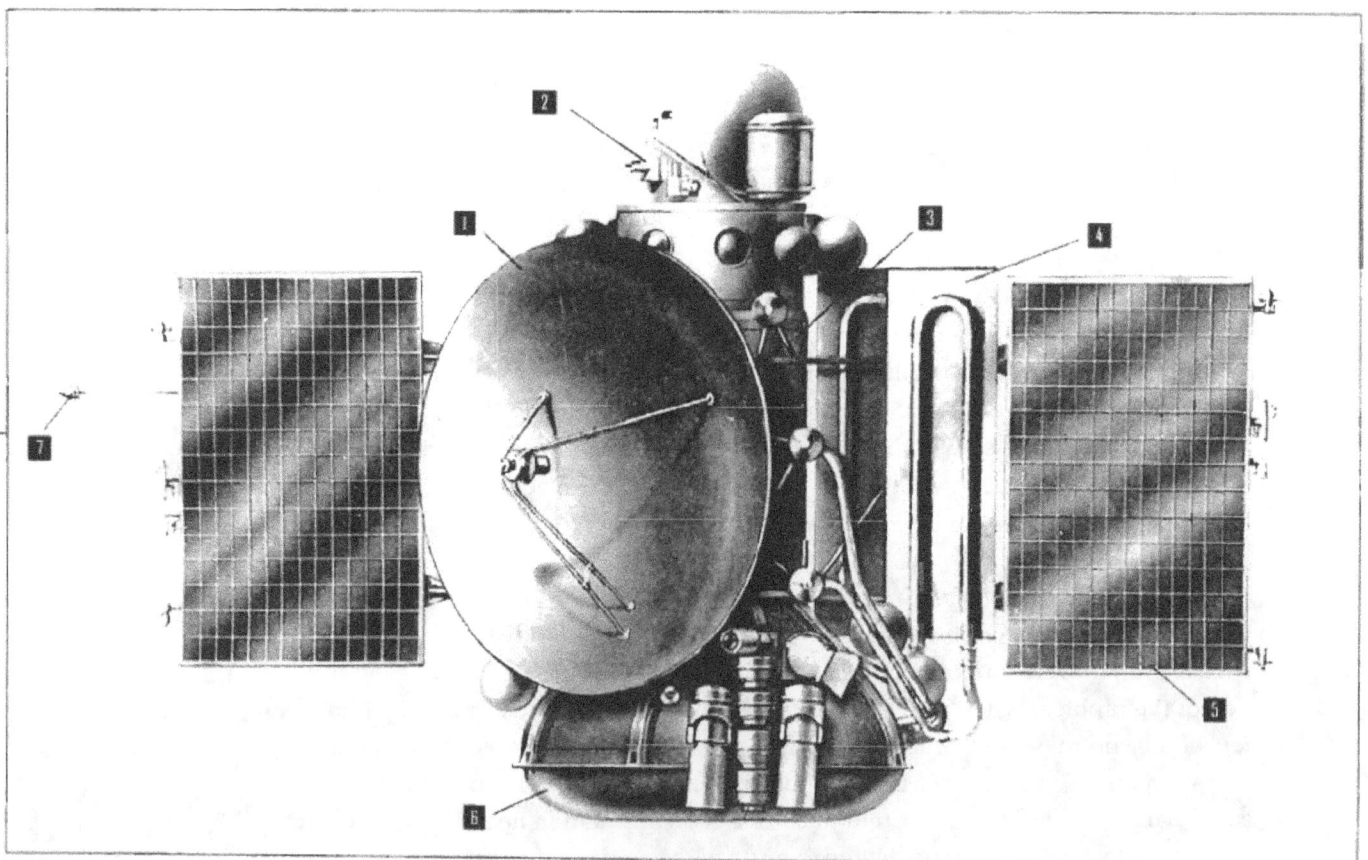

could make this crucial decision. About 5:00 a.m., we arrived at KIS of the Lavochkin design bureau.

All operators were already at their workplaces waiting for our instructions. As usual, the order of A.P. Milovanov was fulfilled. In cooperation with the specialists from NPO AP, we again checked the connections of the joints. They worked well. In an hour and a half, the test was completed. All parameters were in the proper range. Now we could rest. Soon the test of the power system was completed, and the spacecraft was delivered to the launch facility to be prepared for launch.

On June 21 and 25, 1973, the Mars 4 and 5 spacecraft (Figure 18) were launched. On August 5 and 9, 1973, the Mars 6 and 7 spacecraft (Figure 19) were launched. The Mars 4 and 5 spacecraft were launched earlier to have time to be stabilized in Martian satellite orbits and thereafter receive the information from the landers that would be delivered to the Martian surface by the Mars 6 and 7 spacecraft.

The exhausting work of the flight control group was begun. In the beginning, the flight proceeded properly, but after 2 months, the transmission of telemetry information from the Mars 6 spacecraft stopped.

It was the spacecraft whose control system we attempted to save at night. Probably, the 2T-312 transistor had failed. Only the channel for commands and for outbound trajectory measurements remained. Until the end of the flight mission, we did not receive any information about the spacecraft's systems. During the remaining 5 months of the flight, the Mars 6 spacecraft operated completely autonomously and completed intricate programs without commands from Earth.

Approaching Mars, the spacecraft measured its position with respect to Mars, calculated strength and direction of the correction impulse for its trajectory, performed the correction of the trajectory, calculated the direction of the lander descent, separated the lander from the spacecraft, and received and transmitted to Earth information from the lander at the point when the parachute system started to be deployed.

Figure 19.
The Mars 6 and 7 Spacecraft

The silent spacecraft accomplished the flight mission completely. The additional transmission channel was operative only during the descent of the lander. Information from the Martian surface had to be transmitted by the main radio channel to the Mars 5 spacecraft; however, the signal did not reach its destination.

An analysis was performed after the flight and showed that the Mars 6 craft landed in the vicinity of the Valley Samara, which was characterized by a V-shaped cross-section profile. The coordinates of the landing site were 23° 54′ S and 19° 25′ W. Perhaps the landing occurred on a steep hill?

Because of the failure of the onboard systems, the Mars 4 and 7 spacecraft did not accomplish their flight program. In the Mars 4 spacecraft, the correction braking propulsion system failed, and as a result, the spacecraft did not enter a Martian satellite orbit. The Mars 4 spacecraft came within 2,200 kilometers of the Martian surface and made Mars images from its flyby trajectory. After separation from the spacecraft, because of the failure of the 2T-312 transistor, the Mars 7 lander was not put in an encounter trajectory and missed the planet by 1,300 kilometers.

Only the Mars 5 spacecraft went into satellite orbit with the following parameters: the height of apocenter 32,500 km, the height of pericenter 1,760 km, an inclination of 35 degrees to the Martian equator, and a 25-hour period of rotation. The spacecraft accomplished its flight mission completely. Mars 5 high-resolution images provided additional data about the planet (Figure 20).

Figure 20.
Image of the Martian Surface Obtained From the Mars 5 Spacecraft

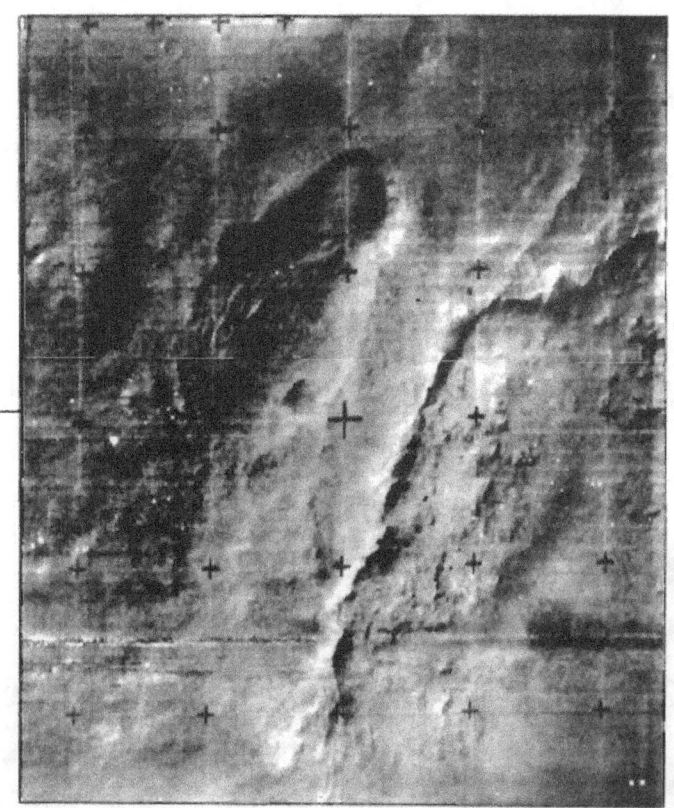

Minister Afanasiev was deeply concerned by the numerous systems malfunctions that led to the failure of the Martian flight mission and to the loss of the Soviet Union as lead in this field. The results of Project M-71 were analyzed at a ministry meeting. Right after this meeting, S.A. Afanasiev and I met A.I. Shokhin, who was the Minister of the Electronic Industry. We asked him only one question: "How do we avoid numerous failures during the next flights?" We were not concerned with a single failure because the spacecraft systems continued to operate in that situation.

A.I. Shokhin offered two suggestions. The first one was to use the electronic components that were designed about 10 years ago and were fabricated in accord with well-developed technology. The second option was to choose advanced electronic components using a special program.

Without a second thought, we rejected the first suggestion. The second option was accepted. Programs were developed to choose the electronic components (they got the abbreviation MB). They increased the reliability of the onboard instruments. On the other hand, the cost of the MB electronic components was almost 10 times the previous cost. But, as they say, "The goal is worth the effort."

6.1 Project 5NM

One of the main missions of the manned Apollo spacecraft was to deliver lunar soil to Earth. This problem could be solved by an automatic spacecraft as well. The idea of delivering lunar soil to Earth by automatic spacecraft was in the air and was awaiting a favorable situation to be realized.

In the summer of 1968, I visited Babakin. In my presence, Yu.D. Volokhov, who was the deputy director of the design bureau, and Babakin discussed the Soviet manned lunar program, which, in their view, was behind schedule. At this time, Babakin spoke about the possible use of the land rover developed at the Lavochkin design bureau for moving the Soviet cosmonauts about the lunar surface to collect samples of lunar rocks and deliver them back to Earth.

A sudden thought went through my mind, and I said, "What if on the lunar landing platform the land rover would be replaced by a rocket. The rocks could be collected and delivered to Earth." Babakin focused on this idea and thought for a while. His eyes moved to the right, then to the left, probably following his thoughts. Eventually, he returned to his usual mood and said, "But it is a very difficult problem." I responded, "I do not have expertise in designing lunar spacecraft and cannot make a decision." However, the idea attracted Babakin, and he issued instructions to work on it.

After careful consideration, it was concluded that the project could be realized only if the advanced technical decisions were used and strict weight limitations taken into account. Babakin gave much attention to this project. He was eager to see the project completed, and he always participated in the solution of intricate scientific and technical problems. Probably, the project was accomplished successfully because of the enormous efforts of Babakin.

At the beginning of 1970, the lunar project's problems had been solved. At this time, Babakin gave instructions for a technical proposal to consider how the

Martian soil would be delivered to Earth. In the summer of 1970, the technical assignments for Project 5NM had been completed. It was planned that in September 1975, the powerful N-1 rocket would launch the spacecraft and inject a payload of 98 tons in a predetermined Earth satellite orbit. This weight included the weight of the 5NM spacecraft (20 tons) (Figure 21). The orbiter, which weighed 3,600 kilograms, was designed to deliver the lander to Mars and to receive the telemetry data from the lander during its descent and landing on the Martian surface. The orbiter included the toroidal instrument module, which came from the M-71 project, and the propulsion system with the spherical fuel tank, which came from the M-69 project.

The lander, which weighed 16 tons, had an aeroshell screen with a central solid area with a diameter of 6.5 meters. At the perimeter of the solid cover, 30 petals were attached. After the spacecraft entered an interplanetary trajectory, the petals were opened, creating an aerodynamic cone with a diameter of 11 meters. The aeroshell screen covered the instrument module, which included the system that controlled the soft landing. Also, this system included the velocity meter, which worked on the Doppler principle, and the altimeter. In addition, the instrument module included the radio system, the program timing device, and the power system.

Figure 21.
The 5NM Spacecraft Designed to Deliver the Samples of the Martian Soil to Earth—
(1) returning capsule, (2) orbital Mars-Earth module, (3) second stage, (4) first stage, (5) landing stage, (6) aeroshield cover, (7) instrument module, (8) orbital Earth-Mars module, and (9) lander

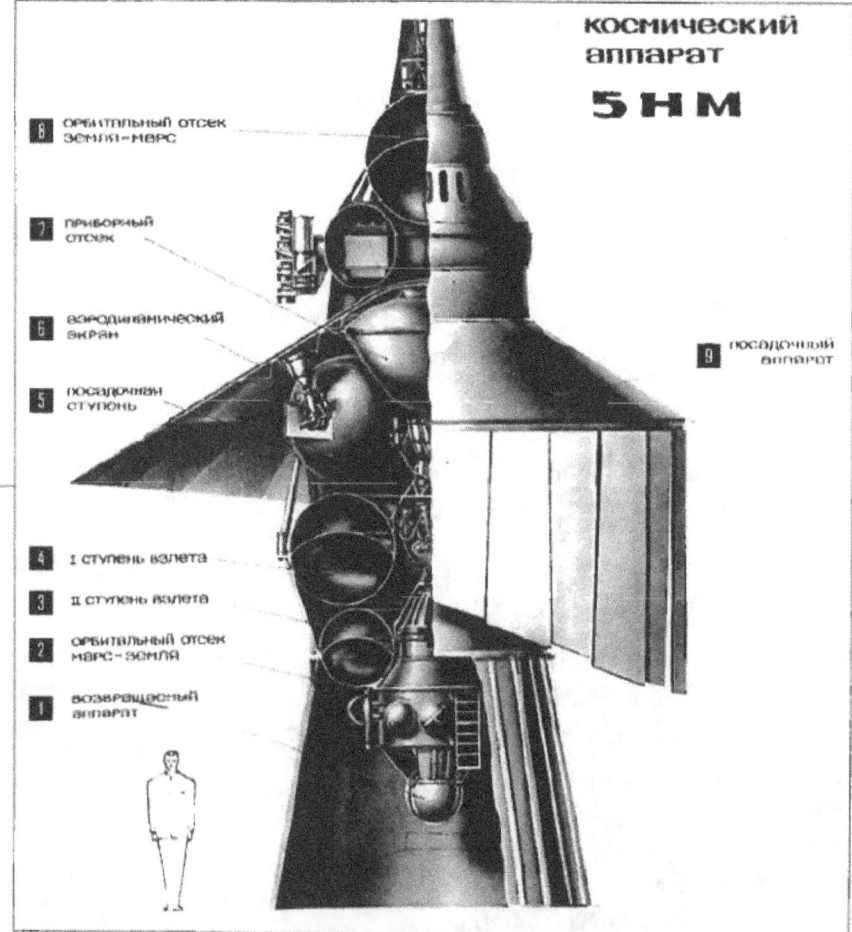

The propulsion soft landing system was composed of four spherical fuel tanks and four propulsion engines, which controlled the thrust. On the top of the propulsion system, a two-staged booster rocket was installed. The rocket included the orbital Mars-Earth module, which weighed 750 kilograms, and a return craft, which weighed 15 kilograms and was designed to deliver 200 grams of Martian soil back to Earth. The design of the Venus 4/6 orbiter was used for fabricating the orbital Mars-Earth module.

To transfer the spacecraft from Earth's orbit to an interplanetary orbit, a two-staged boosting rocket was proposed. When the spacecraft would approach Mars, a trajectory correction had to be performed. Then the lander would be separated from the orbiter. To receive the telemetry data from the lander, the orbiter would be placed in a flyby trajectory.

At the same time, the lander would move along its trajectory and enter the Martian atmosphere. The asymmetrical aeroshield cover would cause the lander to perform a gliding descent. When the lander's speed would reach 200 m/sec, the aeroshield cover would be discarded, the propulsion system would be initiated, and the craft would land on the Martian surface.

When designing this project, I recalled the comments of Korolev at the meeting in the Lavochkin design bureau, who said that landing should be performed by engines without parachutes. Certainly, for this project, it would be beneficial to perform the landing without a parachute. It was planned that after landing communication between the lander and Earth would be maintained in the decimeter band.

Upon a command from Earth, samples of the Martian soil that had been chosen in the panoramic image areas would be collected and loaded in the capsule. In 3 days, after the lander autonomously defined its position by a command from Earth, a booster rocket with the orbital Mars-Earth module and with the returning capsule would be launched in a Martian satellite orbit.

The parameters of the orbit should be the following: a pericenter height of 500 km and a rotation period of 12 hours. After 10 months in a Martian satellite orbit waiting for a favorable date, the orbiter with the capsule would be transferred to an interplanetary trajectory and returned back to Earth (Figure 22). Approaching Earth,

Figure 22.
The Profile of an Earth-Mars-Earth Flight If the Spacecraft Would Have Been Landed in 1975—(1) 30.11.76, spacecraft is positioned behind the Sun, (2) 22.09.76, arrival at Mars, (3) interplanetary Earth-Mars trajectory, T=377 days, (4) Martian orbit, (5) 14.05.78, arrival at Earth, (6) 7.4.77, (7) interplanetary Mars-Earth trajectory, T=291 days, (8) 6.2.77, (9) Earth's orbit, (10) 30.11.76, (11) Sun, (12) 28.09.76, Earth's position at the time of the spacecraft arrival at Mars, (13) 27.07.77, spacecraft is launched to Earth, (14) time that the Martian satellite spent in orbit, T=302 days, (15) 17.09.75, launch to Mars, (16) 27.07.77, position of Earth at the time when the spacecraft was launched from Mars, (17) 7.4.77, (18) duration of the expedition—970 days, and (19) 6.22.77

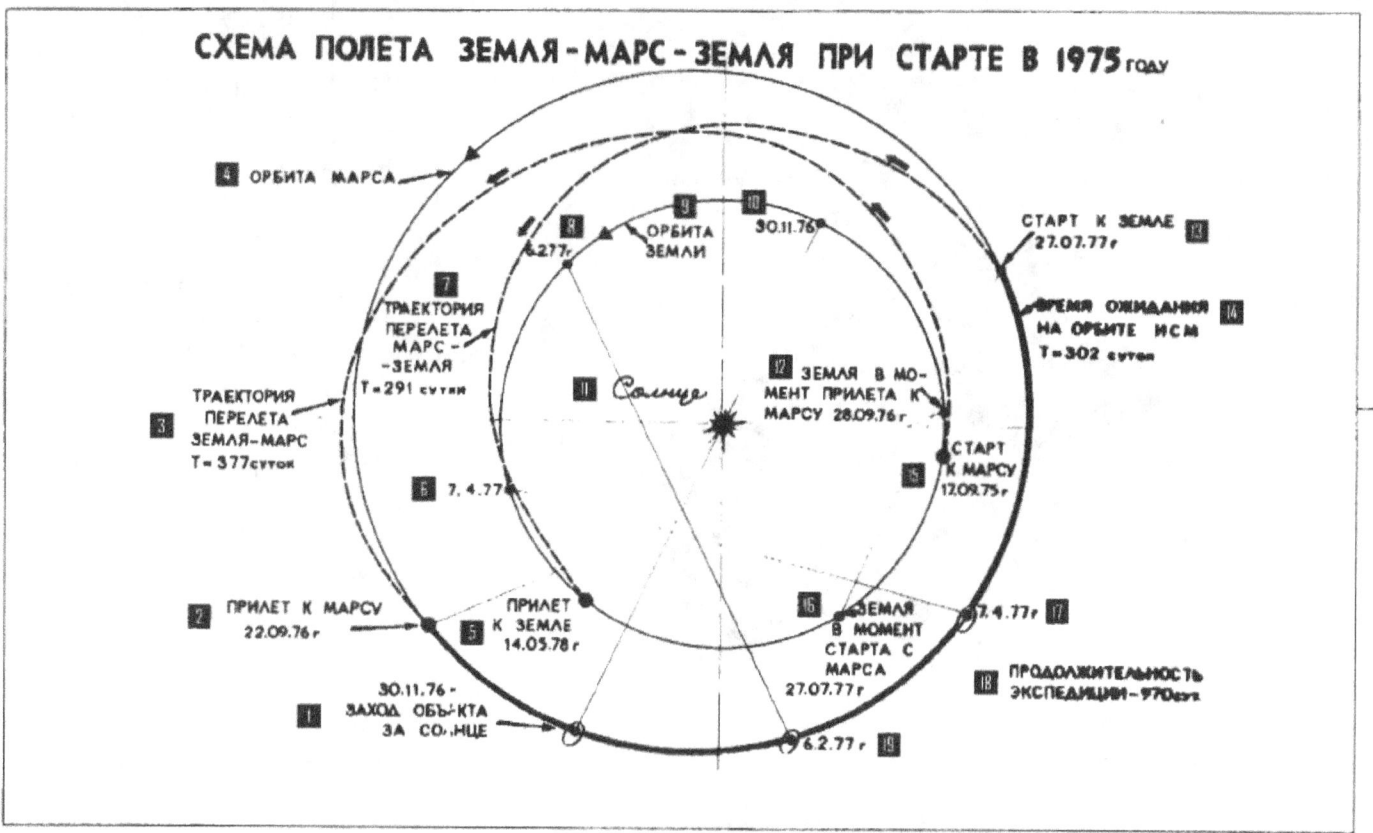

СХЕМА ПОЛЕТА ЗЕМЛЯ – МАРС – ЗЕМЛЯ ПРИ СТАРТЕ В 1975 году

4 ОРБИТА МАРСА

9 10 ОРБИТА ЗЕМЛИ

30.11.76

СТАРТ К ЗЕМЛЕ 13 27.07.77г

8 6.2.77г

7 ТРАЕКТОРИЯ ПЕРЕЛЕТА МАРС – ЗЕМЛЯ T=291 суток

ВРЕМЯ ОЖИДАНИЯ 14 НА ОРБИТЕ ИСМ T=302 суток

3 ТРАЕКТОРИЯ ПЕРЕЛЕТА ЗЕМЛЯ–МАРС T=377суток

11 Солнце

12 ЗЕМЛЯ В МОМЕНТ ПРИЛЕТА К МАРСУ 28.09.76г

СТАРТ К МАРСУ 12.09.75г

6 7.4.77г

2 ПРИЛЕТ К МАРСУ 22.09.76г

5 ПРИЛЕТ К ЗЕМЛЕ 14.05.78г

16 ЗЕМЛЯ В МОМЕНТ СТАРТА С МАРСА 27.07.77г

7.4.77г 17

18 ПРОДОЛЖИТЕЛЬНОСТЬ ЭКСПЕДИЦИИ–970сут

1 30.11.76 - ЗАХОД ОБЪЕКТА ЗА СОЛНЦЕ

6.2.77г 19

the return capsule would be separated from the orbiter. Because of aerodynamic braking, the speed of the capsule would decrease to 200 m/sec. At this time, the parachute system would be deployed and the radar beacon, which facilitates the capsule search, would be activated.

In 1973, to test the spacecraft and lander systems as well as to explore Mars with a rover, it was planned to launch the 4NM spacecraft (Figure 23). The project for the delivery of Martian soil was discussed at a scientific-technical meeting. Eventually, we came to the conclusion that it would not be feasible to accomplish it on schedule for the following reasons:

Figure 23.
The 4NM Spacecraft Designed for Mars Exploration With a Mars Rover—(1) Mars rover, (2) lander, and (3) Earth-Mars orbiter

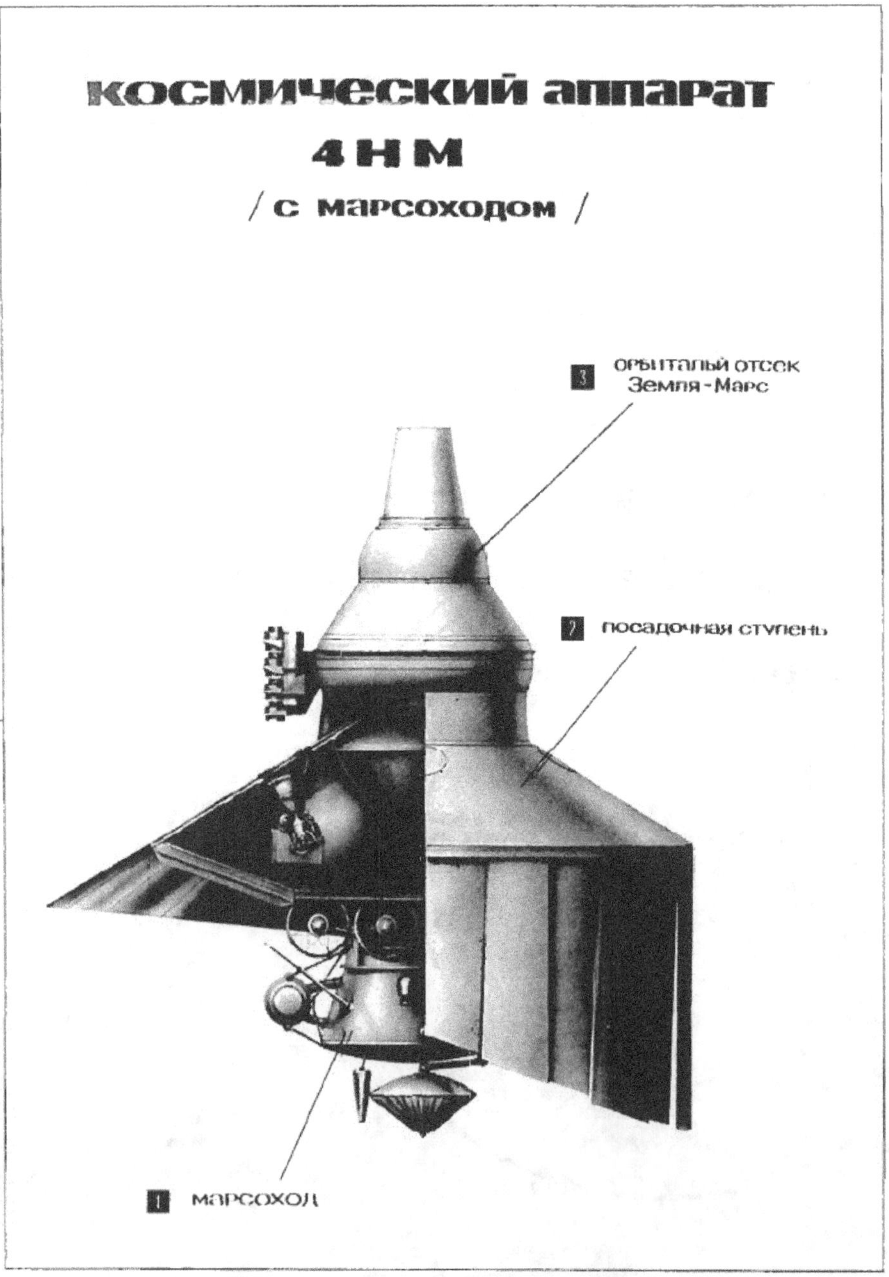

космический аппарат 4 Н М

/ с марсоходом /

3 орбитальный отсек
Земля - Марс

2 посадочная ступень

1 марсоход

1. Biological contamination of Earth was possible. If the parachute system failed, the capsule would brake up. Biologists believed that Martian microbes might be present in Martian soil. On Earth, the Martian microbes could propagate with a very high speed. This experiment could be a major tragedy for Earth.

2. The onboard systems and instruments had not been tested in real flight. Apparently, we were not aware how they would operate during almost 3 years.

However, Minister Afanasiev liked this project. He admitted that our arguments on biological insecurity were reasonable. Simultaneously, being confident that biological problems would be solved soon, he attempted to convince Babakin to begin the project . Nevertheless, Babakin did not agree. The minister suggested that I become the main designer of the project. He promised all kinds of help. The help of such a powerful and influential chief meant a lot, but I could not accept his offer. Afanasiev was disappointed, but he did not forget this project.

6.2 Project 5M

By 1974, we had acquired extensive experience in developing Martian spacecraft and testing their operation in the conditions of real flight. Again, Afanasiev ordered the development of a project for the delivery of Martian soil.

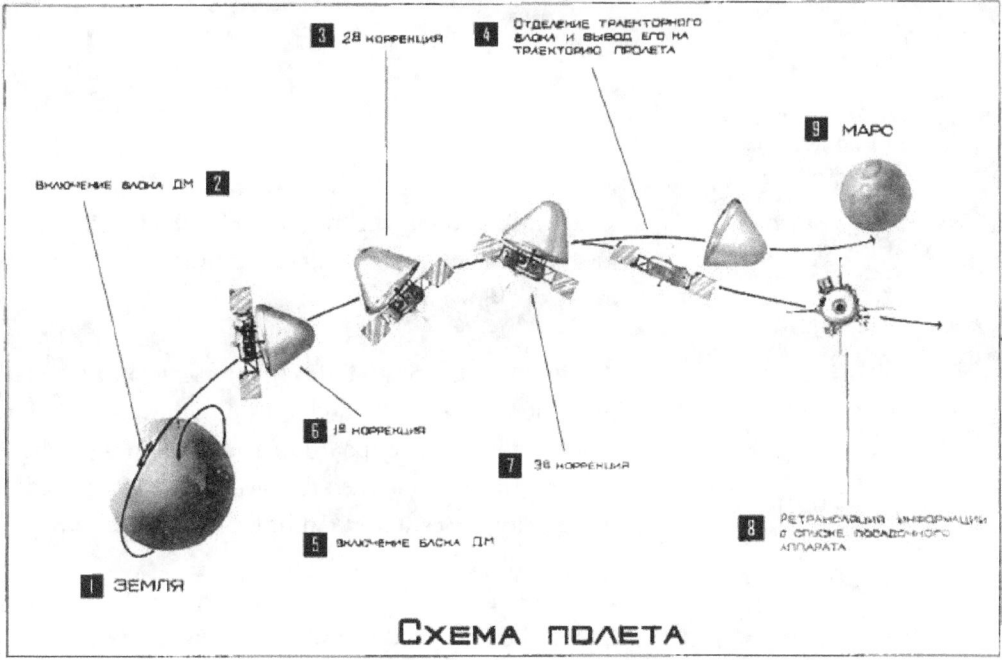

СХЕМА ПОЛЕТА

Figure 24.
The Flight Profile of the 5M Spacecraft — (1) Earth, (2) activation of block DM, (3) second correction, (4) separation of reactive block and its injection into flyby trajectory, (5) insertion of the DM block, (6) first correction, (7) third correction, (8) retransmission of the information on the lander descent, and (9) Mars

By this time, the situation had changed. The fabrication of the N-1 rocket had been stopped. Only the Proton rocket could be used for launching the spacecraft. The Proton rocket was able to place a payload of 22 tons in an Earth orbit. This weight included the booster block D.

This payload was insufficient to deliver Martian soil to Earth. To increase the spacecraft weight, we decided to perform the docking of two payloads launched by Proton rockets in an Earth orbit. Specifically, the first booster block D would be docked with the second booster block D, which in turn would be docked with the spacecraft. After docking and subsequent ignition of the D blocks, the spacecraft, which weighed 8,500 kilograms and consisted of the orbiter and lander, would be placed in an interplanetary trajectory.

The flight had to be performed with a trajectory that would eventually approach Mars. Upon approaching the planet and to receive data transmitted from the lander, the orbiter would be transferred to a flyby trajectory (Figure 24). The lander would perform a gliding descent and land on the Martian surface. Using the panoramic images, there would be a command from Earth and the Martian soil would be collected and loaded in the capsule, which was installed in the second stage of the booster rocket. The rocket, which weighed 2,000 kilograms, had to deliver the capsule with the soil to a Martian orbit.

To deliver the capsule back to Earth, the spacecraft with the return craft should be placed in a Martian orbit. In this orbit, the spacecraft had to be docked with the last stage of the rocket that contained the capsule. Then the capsule should be reloaded in the return craft.

When a favorable starting date arrives, the soil should start its trip to Earth. To place the spacecraft in a Martian orbit, it was necessary to launch a third Proton rocket. To avoid biological contamination, it was proposed to place the return craft in an Earth orbit, dock it with a manned spacecraft, reload the return craft into the manned

Figure 25.

The Design of the Descent and Landing of the 4M Lander Vehicle –(1) entrance in the atmosphere, altitude, H=100 km, speed of the entrance in the atmosphere, V=5.6 km/sec, angle of entrance in the atmosphere, θ=–13°, (2) rotation of the lander around its axis to change the direction of G-force, H=18.1 km, V=4.75 km/sec, θ=12.6°, (3) closest approach of the lander to the Martian surface, H=5.5 km, V=1.0 km/sec, (4) landing, predetermined conditions: V vertical ≤3 m/sec, V horizontal ≤1 m/sec, (5) beginning of the aerodynamic braking, H=55.9 km, V=5.62 km/sec, θ=–10.9°, (6) activation of the instruments that work on the Doppler principle, H=9.4 km, V=700 m/sec, θ=2.5°, (7) insertion of the engines of the precision braking H=3.03 km, V=355 m/sec, θ=26°, (8) insertion of the engine of the main braking, H=213 km, V=338 m/sec, θ=–29.4°, (9) dropping of the engine container, H=1.85 km, V=50.8 m/sec, θ=–49.4°, (10) descent with constant speed, H=10–30 m, removing of the undercarriage and dropping the aeroshell cover, (11) maximum overloading, H=10.5 km, V=3.58 km/sec, θ=–12.6°, (12) Mars rover activation, (13) nominal trajectory of the lander, (14) altitude (km), and (15) distance (km)

spacecraft, and deliver the soil to Earth. Therefore, to deliver the Martian soil to Earth, it was necessary to launch three Proton rockets and to perform three automatic dockings in the space.

Apparently, this project was too intricate and unreliable. Because of the strict weight limitations, it was planned to use modern onboard instruments. Similar to Project 5NM, to check the onboard instruments in the conditions of real flight, it was planned at first to launch the 4M spacecraft to study Mars with a Mars rover. The designs for descent and landing for both crafts were identical (Figure 25).

Unfortunately, at this time, cooperation between the Soviet Union and the United States in the field of Mars exploration had not started. After the American Apollo program was completed, the Saturn V rockets, which were capable of placing a payload of 138 tons in an Earth's orbit, were destroyed. There was no doubt that if our countries combined their efforts and used a Saturn V rocket, the problem of delivering the Martian soil to Earth would be solved and complete biological security of Earth would be provided.

I presented the results of our work to main designer Kryukov and simultaneously emphasized the complexity of the project and the low probability of its success. Eventually, I suggested postponing the project until a better time. Kryukov did not agree with me, but I could not continue to work on the project, which, in my view, had no chance of complete success. As a result of this disagreement, the future development of the project was given to V.P. Panteleev, the deputy of the main

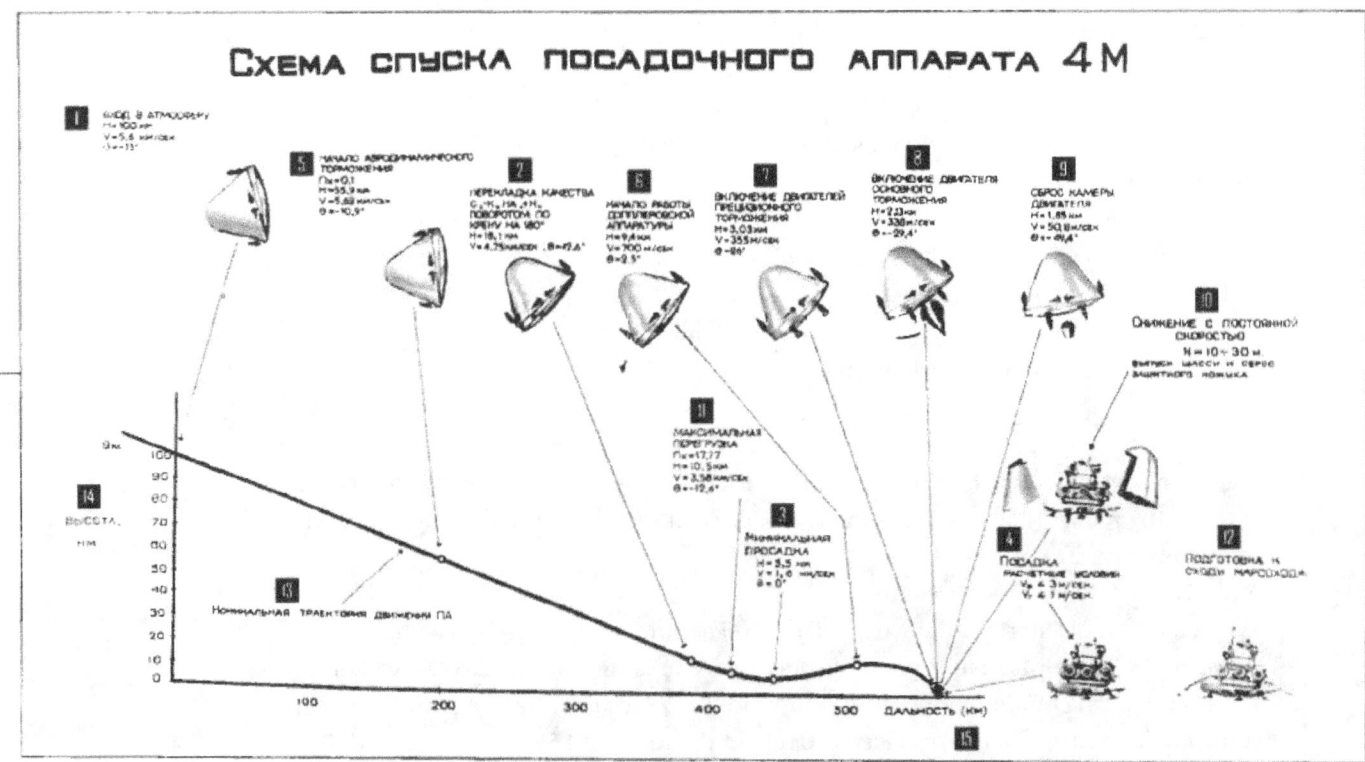

designer. V.P. Panteleev was a highly qualified specialist and was persistent in achieving his goals.

6.3 The Continuation and the End of the 5M Project

To simplify the flight design and to decrease the number of space dockings, V.P. Panteleev decided to increase the weight of the spacecraft after modifying the booster blocks D. This modification included the installation of additional fuel lines and their connectors to transfer fuel after the blocks were docked. The fuel had to be transferred from an active block D, which functioned as a first stage, to a passive block D, which functioned as the second stage.

Both blocks injected the spacecraft in an interplanetary orbit. Because of these modifications, the weight of the spacecraft was increased from 8,500 to 9,335 kilograms. The additional increase of the spacecraft weight required a change in the design of the lander's descent. To achieve this goal, the lander's gliding descent had to be replaced by a ballistic descent. In addition, the lander's shape and design had to be changed.

At the first stage of the project, the lander was designed in the shape of a headlight. In this stage, the headlight was replaced with a conical aeroshield cover, which was like an umbrella and had a diameter of 11.35 meters. The center was a solid part of the cover with a diameter of 3 meters, into which the thick, tube-like beryllium spokes were installed. The spokes were covered with fiberglass. Before the spacecraft was launched, the spokes were folded along the lander's axis. After the spacecraft would be injected into an interplanetary trajectory, the spokes would be opened to create the aeroshield cover.

All suggestions about a possible weight decrease of onboard instruments had been carefully analyzed. As a result, the design of the spacecraft, which weighed 9,135 kilograms and consisted of the orbiter (1,680 kilograms) and lander (7,455 kilograms), was developed. The lander included a two-stage booster rocket, which weighed 3,190 kilograms, and the Mars-Earth return craft.

In January 1976, Kryukov approved the preliminary design of the return craft, which had to deliver the Martian soil and whose weight should not exceed 2 percent of the spacecraft weight. With these weight limitations, there was hope that the project could be completed. However, the problem of biological contamination had not been solved. Simultaneously with the development of the technical assignment, we started the second stage of the project to find a solution to the weight limitations. To do that, a number of highly advanced technical solutions to decrease the weight of the spacecraft were implemented.

The idea of Academician A.P. Vinogradov was very helpful. He suggested conducting thermal sterilization of the soil samples while in the Martian orbit. This suggestion completely solved the problem of biological security of Earth. In addition, the implementation of this idea would lead to a significant decrease in the weight of the spacecraft. Apparently, when the weight of the sample return craft would be decreased by 1 kilogram, the weight of the spacecraft could be decreased by 10 kilograms. The following parts were removed from the return craft: parachute, radar, battery, and automatic unit. The return craft, with a weight of 7.8 kilograms and speed of 12 km/sec, would enter Earth's atmosphere like a meteorite. After aerodynamic braking, the speed of the craft would be decreased to a few tens of meters per second, and the craft would fall to Earth.

According to calculations, the scatter of the craft landing site was restricted to a circle with a radius of 40 kilometers. The search for the craft was to be performed with helicopters equipped with instruments designed to detect a radioactive source in the returning craft. The second stage of the project provided the opportunity to decrease the weight of the spacecraft up to 4 percent of the spacecraft weight.

Many organizations were involved in the development and fabrication of the spacecraft. In 1978, when the first models and parts of the spacecraft were designed, the Head Institute issued a statement in which the complexity of the project, its high cost, and low probability of success were indicated. Based on this statement, Minister Afanasiev decided to discontinue the project for Martian soil return.

This decision seriously damaged the prestige of the Lavochkin design bureau. Apparently, many organizations strongly believed in the ability of the Lavochkin design bureau to accomplish the most intricate of technical assignments. This was the reason their specialists concentrated on this project. But all of a sudden, they were out of business!

The main designer, Kryukov, who said he was guilty of mismanaging the project, resigned. At that time, V.M. Kovtunenko, the deputy to the main designer of Yushnoe NPO in Dnepropetrovsk, was appointed the main designer of the Lavochkin design bureau.

Kovtunenko had been working at Yushnoe NPO in Dnepropetrovsk for many years. In the beginning, he developed ballistic rockets. During the last 10 years, before he was appointed the main designer of the Lavochkin design bureau, Kovtunenko was in charge of the Intercosmos satellites, which were developed according to a cooperative program among socialist countries and were designed to study the space near Earth. The Intercosmos satellites were developed using the design of one of the Cosmos series satellites. They had small weight, a simple design, and were launched in an Earth's orbit with the two-stage Tsyklon rocket.

Naturally, the problems that had to be solved by the recently appointed main designer to develop the interplanetary automatic spacecraft were not similar to the problems with which he dealt while developing the Intercosmos satellites. Kovtunenko decided to find his own way in this new field. He was fascinated with the idea of creating the multipurpose spacecraft to study the Moon, Venus, and Mars. In 1979, the design of the UMVL spacecraft (Universal Mars, Venus, Luna) started to be developed.

At the same time, Kovtunenko decided to continue the development of the Venus automatic spacecraft initiated by main designer Kryukov. These spacecraft had to be launched in the next 3–4 years. During this period of time, Kovtunenko hoped to develop the UMVL spacecraft and use it to perform the broad planetary and lunar study.

However, the pace of the development of the new spacecraft was slow. That could be explained by an unlucky choice of the project manager, who was the former party functionary, a skillful politician whose only desire was to be recognized and nothing more. In addition, Kovtunenko had a difficult time in establishing effective contacts with such "heavyweights" of the space industry as V.P. Glushko, N.A. Pilyugin, M.S. Ryasanskiy, and others. Apparently, without their support, it was difficult to achieve a quick success in the development of new spacecraft. Therefore, the Phobos 1 and 2 spacecraft (Figure 26) were launched only in 1989, 10 years after their development had been initiated. The flight program outlined: (1) to place the Phobos 1 and 2 spacecraft in a Martian orbit, (2) to approach the Martian satellite Phobos, (3) to land two automatic stations on Phobos' surface, (4) to conduct a remote distance study of Phobos soil chemistry from a flyby trajectory, and (5) to conduct an extensive study of Mars from its satellite orbit.

Figure 26.
The Phobos 2 Spacecraft

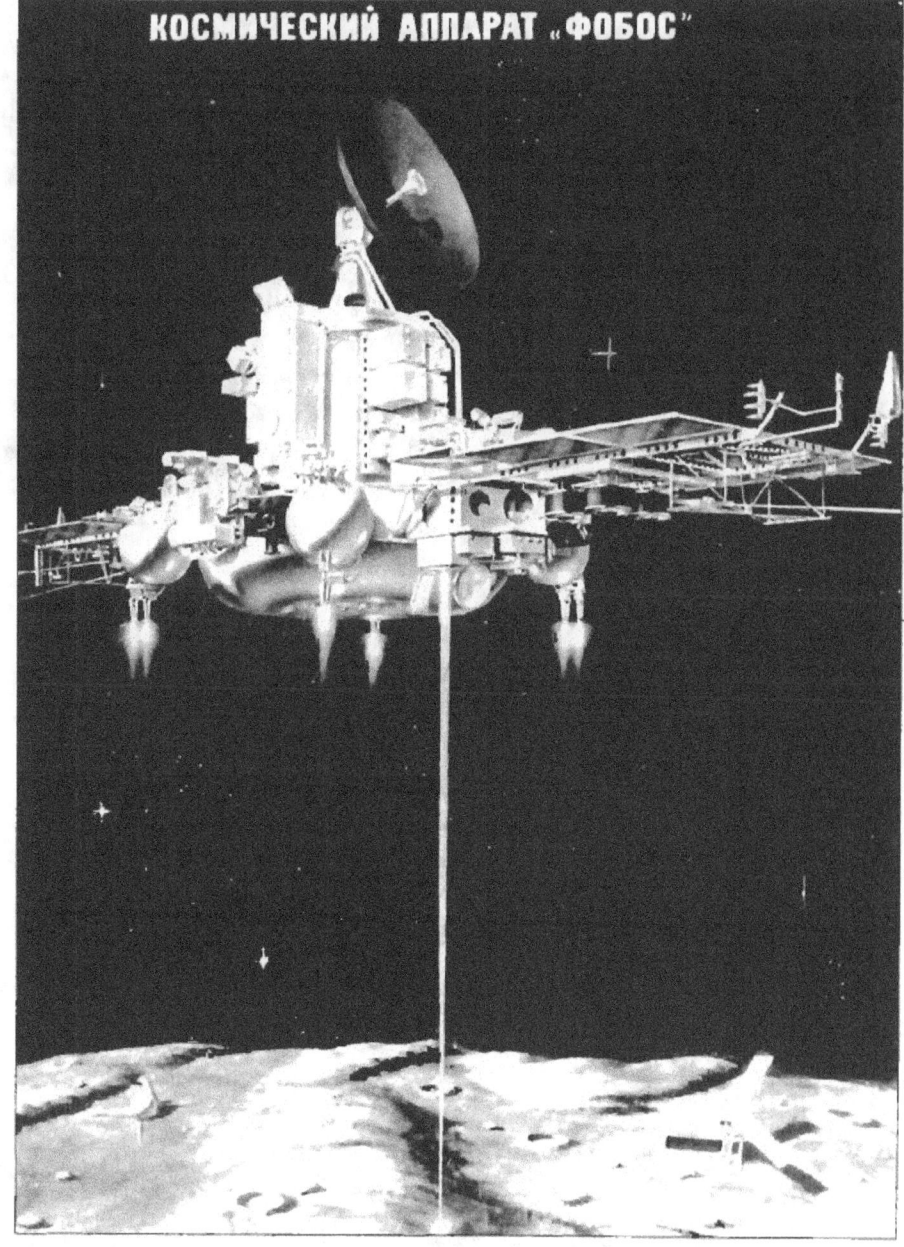

КОСМИЧЕСКИЙ АППАРАТ «ФОБОС»

МЕЖДУНАРОДНЫЙ ПРОЕКТ «МАРС»

НПОим.С.А.ЛАВОЧКИНА

ГЕОХИ РАН

ВНЕДРЯЕМЫЙ ЗОНД

Figure 27.
The Mars 96 Spacecraft

Unfortunately, neither spacecraft accomplished the flight program because of the huge mistakes that were made during their construction and because of neglecting the rules established in Korolev OKB when designing the first interplanetary spacecraft.

The first spacecraft failed in the interplanetary trajectory as a result of two unfavorable circumstances: (1) the failure to issue onboard the correct command to transfer the propulsive mass to the micro-engines of the spacecraft's attitude system and (2) a logic error in the onboard prohibition systems that should have disconnected the fuel line while the spacecraft was in a position of three-axis stabilization.

The second spacecraft failed in a Martian satellite orbit. That happened because the computer program named "e minimal" was not included in the onboard software. If the power voltage decreased below a predetermined level, the program "e minimal" would have automatically issued a command to disconnect all onboard systems except the systems needed for the spacecraft to survive. All interplanetary automatic spacecraft developed at OKB-1 and the Lavochkin design bureau carried the program "e minimal" onboard, and the first attempt to exclude it from the computer software led to sad result.

The Mars 96 spacecraft (Figure 27) failed as well. This spacecraft was built to perform an extensive study of Mars from its orbit with penetrators dropped to the Martian surface. The Mars 96 spacecraft was not injected in an interplanetary trajectory.

Today, it seems that the Soviet and, thereafter the Russian, program of Mars exploration was ended on a sour note. However, there is the Russian proverb that says: "One beaten person is worth two unbeaten ones."

As soon as the Russian economy is stabilized, young creative minds who have already developed the original approach to Mars and Phobos exploration will overcome and succeed.

INDEX

Launius, Roger D., and Gillette, Aaron K. Compilers. *The Space Shuttle: An Annotated Bibliography* (Monographs in Aerospace History, No. 1, 1992).

Launius, Roger D., and Hunley, J.D. Compilers. *An Annotated Bibliography of the Apollo Program* (Monographs in Aerospace History, No. 2, 1994).

Launius, Roger D. *Apollo: A Retrospective Analysis* (Monographs in Aerospace History, No. 3, 1994).

Hansen, James R. *Enchanted Rendezvous: John C. Houbolt and the Genesis of the Lunar-Orbit Rendezvous Concept* (Monographs in Aerospace History, No. 4, 1995).

Gorn, Michael H. *Hugh L. Dryden's Career in Aviation and Space* (Monographs in Aerospace History, No. 5, 1996).

Powers, Sheryll Goecke. *Women in Aeronautical Engineering at the Dryden Flight Research Center, 1946–1994* (Monographs in Aerospace History, No. 6, 1997).

Portree, David S.F., and Trevino, Robert C. Compilers. *Walking to Olympus: A Chronology of Extravehicular Activity (EVA)* (Monographs in Aerospace History, No. 7, 1997).

Logsdon, John M. Moderator. *The Legislative Origins of the National Aeronautics and Space Act of 1958: Proceedings of an Oral History Workshop* (Monographs in Aerospace History, No. 8, 1998).

Rumerman, Judy A. Compiler. *U.S. Human Spaceflight: A Record of Achievement, 1961–1998* (Monographs in Aerospace History, No. 9, 1998).

Portree, David S.F. *NASA's Origins and the Dawn of the Space Age* (Monographs in Aerospace History, No. 10, 1998).

Logsdon, John M. *Together in Orbit: The Origins of International Cooperation in the Space Station Program* (Monographs in Aerospace History, No. 11, 1998).

Phillips, W. Hewitt. *Journey in Aeronautical Research: A Career at NASA Langley Research Center* (Monographs in Aerospace History, No. 12, 1998).

Braslow, Albert L. *A History of Suction-Type Laminar-Flow Control with Emphasis on Flight Research* (Monographs in Aerospace History, No. 13, 1999).

Logsdon, John M. Moderator. *Managing the Moon Program: Lessons Learned from Project Apollo* (Monographs in Aerospace History, No. 14, 1999).

Perminov, V.G. *The Difficult Road to Mars: A Brief History of Mars Exploration in the Soviet Union* (Monographs in Aerospace History, No. 15, 1999).

Those monographs still in print are available free of charge from the NASA History Division, Code ZH, NASA Headquarters, Washington, DC 20546. Please enclosed a self-addressed 9x12" envelope stamped for 15 ounces for these items.